Globalization of Water

Globalization of Water

Sharing the Planet's Freshwater Resources

By
Arjen Y. Hoekstra and
Ashok K. Chapagain

Blackwell
Publishing

BLACKWELL PUBLISHING
350 Main Street, Malden, MA 02148–5020, USA
9600 Garsington Road, Oxford OX4 2DQ, UK
550 Swanston Street, Carlton, Victoria 3053, Australia

The right of Arjen Y. Hoekstra and Ashok K. Chapagain to be identified as the authors
of this work has been asserted in accordance with the UK Copyright, Designs, and
Patents Act 1988.

Designations used by companies to distinguish their products are often claimed as
trademarks. All brand names and product names used in this book are trade names, service
marks, trademarks, or registered trademarks of their respective owners. The publisher is not
associated with any product or vendor mentioned in this book.

This publication is designed to provide accurate and authoritative information in regard to
the subject matter covered. It is sold on the understanding that the publisher is not engaged
in rendering professional services. If professional advice or other expert assistance is
required, the services of a competent professional should be sought.

First published 2008 by Blackwell Publishing Ltd

2   2008

Library of Congress Cataloging-in-Publication Data

Hoekstra, Arjen Y., 1967–
   Globalization of water: sharing the planet's freshwater resources / by Arjen Y. Hoekstra
and Ashok K. Chapagain.
      p. cm.
   Includes bibliographical references and index.
   ISBN 978-1-4051-6335-4 (hardcover : alk. paper)
 1. Water-supply—Management. 2. Water resources development. 3. Freshwater ecology.
I. Chapagain, Ashok K. II. Title.
   TD345.H54 2008
   333.91—dc22

                                                          2007025137

A catalogue record for this title is available from the British Library.

Set in 10.5/13.5pt Sabon
by SPi Publisher Services, Pondicherry, India.
Printed and bound in Singapore
by Markono Print Media Pte Ltd

The publisher's policy is to use permanent paper from mills that operate a sustainable
forestry policy, and which has been manufactured from pulp processed using acid-free and
elementary chlorine-free practices. Furthermore, the publisher ensures that the text paper
and cover board used have met acceptable environmental accreditation standards.

For further information on
Blackwell Publishing, visit our website at
www.blackwellpublishing.com

Contents

List of Maps

The plates will be found between pages 84 and 85.

Preface

It looks as if the same politicians, academics, and activists who from the late 1980s gathered around the topic of "sustainable development" have since the late 1990s started to organize themselves around the topic of "globalization." Many of the concerns in the debate about sustainability remain valid in the current discourse about globalization. Major themes are still the balance between economic growth and preserving our environment, security of livelihoods, and equity among people and generations. The new element in the current discourse on globalization is the recognition that the ever-increasing material and cultural exchange between people in different parts of the world and the growing mobility of business make sustainable development a true global challenge. In the past few years thousands of papers and hundreds of books have been written about globalization (Lechner and Boli, 2004). The current book focuses on the effects of globalization on water resources management, a topic that has surprisingly not been much addressed before. This is the first book on the subject. It is true that many valuable books have been published about so-called "global water problems," but the term "global" in these books essentially refers to the fact that the problems described occur all over the world. "Global" is used in these publications in the meaning of "widespread." Problems of water scarcity, water pollution, and flooding are indeed common. However, previously available texts have described the problems in essence from a local, national, or river basin perspective. By contrast, the current volume shows that water problems are often caused by mechanisms that can be understood only at a level far beyond that of the river basin. We will show that local water depletion and pollution are often closely linked to the structure of the global economy. We are convinced that many of today's water problems cannot be solved at river basin level, because they are inextricably bound up with the processes that determine where in the world agricultural and industrial production take place and with the written and unwritten rules of global trade. We hope that this book contributes to the reader's understanding of how wise use of water is linked up with how we organize our global society.

We started our research on the "globalization of water" in 2002 and along the way have received help from many of our students. We would like to thank Pham Quoc Hung from Vietnam for his explorative work on quantifying world trade in water in virtual form. We thank Anat Yegnes-Botzer from Israel for carrying out an interesting case study for Israel, and Zhang Dunquiang and Jing Ma, both from China, for doing two different case studies on virtual-water transfer between provinces within China. We would like to thank Xiuying Dong from China and Mesfin Mergia Mekonnen from Ethiopia for their work on developing a computer tool to assess one's individual water footprint. We also thank Abbas Badawi Ashmage Iglal from Sudan and Thewodros Mulugeta Gebre from Ethiopia who carried out joint research on current and future virtual-water flows in the Nile basin. Finally, we thank Rajani Gautam from Nepal for her valuable study on cotton.

We are grateful to all the experts present at the productive International Expert Meeting on Virtual Water Trade held at UNESCO-IHE in the Netherlands in December 2002 (Hoekstra, 2003). We would like to mention in particular Tony Allan, professor at the School of Oriental and African Studies in London, who invented the term "virtual water" and whose work inspired us to explore this field. We are also grateful to Huub Savenije, professor at UNESCO-IHE and the Delft University of Technology in the Netherlands, who has been one of the few who have seen from the beginning that "globalization of water" will become an important theme and has supported us throughout our work with his stimulating ideas.

We thank the National Institute for Public Health and the Environment in the Netherlands for providing financial support for part of our research for this book. We would like to thank in particular Ton Bresser, who has shown a continuing interest in our work. We are grateful to Oxfam Novib for sponsoring the case study on coffee and tea. Finally, we thank the UNESCO-IHE Institute for Water Education and University of Twente for facilitating the research.

When drafting this book we have made use of a number of our earlier publications. Part of the value of this book is that it brings together all the disparate publications into one coherent structure. When writing Chapters 2, 3, and 5 we have drawn most heavily on a report published by UNESCO-IHE (Chapagain and Hoekstra, 2004), Chapagain's PhD thesis (Chapagain, 2006), and two papers, published in *Water Resources Management* (Hoekstra and Chapagain, 2007a) and *Water International* (Chapagain and Hoekstra, 2007a). Chapter 4 builds on a paper that appeared in *Hydrology and Earth System Sciences* (Chapagain et al., 2006a). Chapter 6 is based on a paper presented at a water conference on the occasion of the 400th anniversary of relations between Morocco and the Netherlands, held in Marrakech in November 2005 (Hoekstra and Chapagain, 2007b). Chapter 7 on China builds on an article with Jing Ma (Ma et al., 2006) published in *Philosophical Transactions of the Royal Society of London*, which as we learned is the world's longest-running scientific journal, having appeared since March 1665. Chapter 8 on coffee and tea is based on a paper published in *Ecological Economics* (Chapagain and Hoekstra, 2007b). Chapter 9 on cotton draws upon another paper published in *Ecological Economics* (Chapagain et al., 2006b). Finally, Chapters 10 and 11 are based on a paper presented at a meeting of the Global Water System Project in Bonn in June 2006 (Hoekstra, 2006). Unless mentioned otherwise, the data presented in this book refer to averages for the period 1997–2001.

Arjen Y. Hoekstra
Enschede, The Netherlands

Ashok K. Chapagain
Kathmandu, Nepal

Chapter 1

Introduction

In the world of today, people in Japan indirectly affect the hydrological system in the USA and people in Europe indirectly impact on the regional water systems in Brazil. When you ask somebody how this can happen, the reply will most probably be: through climate change. This answer is likely because much has been reported about the expected effects of past and ongoing local emissions of greenhouse gases on future global temperature, evaporation, and precipitation patterns. Most people are aware that local emissions of greenhouse gases contribute to global climate change and can thus indirectly affect other locations. Little is known, however, about a second mechanism through which people affect water systems in other parts of the world. This second mechanism, which is as "invisible" as climate change but which is today already much more significant, is global trade. International trade in agricultural and industrial commodities creates a link between the demand for water-intensive commodities (notably crops) in countries like Japan, Italy, Germany, and the UK and the water use for production of these commodities in countries such as the USA and Brazil. Water use for producing export commodities for the global market significantly contributes to changes in local water systems. By buying crop products imported from the USA, Japanese consumers put pressure on the water resources of the latter, contributing to the mining of aquifers and emptying of rivers in North America. Well-documented examples are the mined Ogallala Aquifer and emptied Colorado River. European consumers contribute to a significant degree to the water demand in Brazil by buying water-intensive crop and livestock products imported from this country. Well known is the ongoing deforestation of the Amazonian rainforest with its implications for biodiversity, erosion, and runoff.

Though consensus seems to exist that the river basin is the appropriate unit for analyzing freshwater availability and use, in this book we argue that it is becoming increasingly important to put freshwater issues in a global context. Although other authors have already argued thus (Postel et al., 1996; Vörösmarty et al., 2000), we add a new dimension to the argument. International trade in commodities implies long-distance transfers of water in virtual form, where virtual water is understood as the volume of water that has been used to produce a commodity and that is thus virtually embedded in it (Allan, 1998b). Knowledge about the virtual-water flows entering and leaving a country can cast a completely new light on the actual water scarcity of a country. For example, Jordan imports about 5 to 7 billion m^3 of virtual water per year, which is in sharp contrast with the 1 billion m^3 of water withdrawn annually from domestic water sources (Haddadin, 2003; Chapagain and Hoekstra, 2004). This means that people in Jordan apparently survive owing to the import of water-intensive commodities from elsewhere, for example the USA. Jordan's water shortage is largely covered up by intelligent trade: export of goods and services that require little water and import of products that need a lot of water. The positive side of Jordan's trade balance is that it preserves the scarce domestic water resources; the negative side is that the people are heavily "water dependent." A different case is Egypt, a country which has not been willing to become water dependent and in which water self-sufficiency is high on the political agenda. However, with a total water withdrawal inside the country of 65 billion m^3/yr, Egypt still has an estimated net virtual-water import of 10 to 20 billion m^3/yr (Yang and Zehnder, 2002; Zimmer and Renault, 2003; Chapagain and Hoekstra, 2004). This means that even Egypt's water balance is not immune to its pattern of international trade. In fact, there exist no countries in the world where the pattern of trade does not influence the pattern of domestic water use. Developing national water policies without explicitly considering the implications of international trade thus seems to be injudicious. Nevertheless, this is the common practice. In addition, formulating foreign trade policies in water-scarce countries without explicit consideration of domestic water resource availability would seem to be inadvisable as well. Yet, this is what generally happens.

In this book we address questions such as: Is it efficient to import water in virtual form if domestic water resources are scarce? And what are the implications of importing virtual water in terms of the resulting "water dependency"? But once we enter the area of "water and international trade" other questions also arise. If water-intensive products are imported from a distant location, the negative impacts of water use in the area of production will remain invisible for the consumer. Together with the fact that usually only a small fraction of the full cost of water use is included in the price of products, there is little incentive for consumers to change their consumption behavior or otherwise contribute to the mitigation of distant water problems. Thus new questions come up: What are the invisible tele-connections between intensive consumption of water-intensive products in some places on earth and the impacts of water use in other places? And does international trade in water-intensive commodities contribute to the unrestricted growth of consumption of water-intensive products in a world where water becomes increasingly scarce? In order to address these sorts of questions, we use in this book a number of novel concepts such as the "virtual-water content" of a commodity, the "water footprint" of a nation, and the "water saving" as a result of international trade. Let us introduce and explain these concepts one by one.

The virtual-water concept was introduced by Allan (1998a,b, 1999a,b, 2001) when he studied the possibility of importing virtual water (as opposed to real water) as a partial solution to problems of water scarcity in the Middle East. Allan elaborated the idea of using virtual-water import (coming along with food imports) as a tool to release the pressure on scarcely available domestic water resources. Virtual-water import thus becomes an alternative water source, alongside endogenous water sources. Imported virtual water has therefore also been called "exogenous water" (Haddadin, 2003).

The water-footprint concept was introduced by Hoekstra and Hung (2002) when they were looking for an indicator that could map the impact of human consumption on global freshwater resources. The concept was subsequently elaborated by the authors of this book (Chapagain and Hoekstra, 2004; Hoekstra and Chapagain, 2007a). The water footprint shows water use related to *consumption* within a nation, while the traditional indicator shows water use in relation to *production* within a nation. Traditionally, national water use has been measured as the total freshwater withdrawal for the various sectors of the economy. By contrast, the water footprint shows not only freshwater use within the country considered, but also freshwater use outside the country's borders. It refers to all forms of freshwater use that contribute to the production of goods and services consumed by the inhabitants of a certain country. The water footprint of the Dutch community, for example, also refers to the use of water for rice production in Thailand (insofar as the rice is exported to the Netherlands for consumption there). Conversely, the water footprint of a nation excludes water that is used within the national territory for producing commodities for export, which are consumed elsewhere.

The water footprint of a nation consists of three components: blue, green, and gray components. The terms blue and green refer to the source of the water (Falkenmark, 2003). Green water use refers to the use of rainwater, while blue water use refers to use of ground- or surface water. Rain-fed agriculture is fully based on green water, while irrigated agriculture is based on a combination of green and blue water. The industrial and domestic sectors are generally fully based on blue water. The blue water footprint of a nation is the volume of freshwater that evaporated from global blue water resources (ground- and surface water) to produce the goods and services consumed by its inhabitants. The green water footprint is the volume of water evaporated from global green water resources (rainwater stored in the soil as soil moisture). We have expanded the water-footprint concept by including a third form of water use: water use as a result of pollution. We have proposed to quantify this "gray" water footprint by estimating the volume of water needed to dilute a certain amount of pollution such that it meets ambient water quality standards. We have elaborated this idea in the cotton study discussed in Chapter 9.

Active promotion of the import of virtual water in water-scarce countries is based on the idea that a nation can preserve its domestic water resources by importing a water-intensive product instead of producing it domestically. Import of virtual water thus leads to a "national water saving." In addition to this, Oki and Kanae (2004) introduced the idea of a "global water saving." International trade can save water globally when a water-intensive commodity is traded from an area where it is produced with high water productivity (low water input per unit of output) to an area with lower water productivity (high water input per unit of output). Conversely, there can be a "global water loss" if a water-intensive commodity is traded from an area with low water productivity to one with high water productivity. All studies of global water savings and losses as a result of international trade that have been carried out so far indicate that the net effect of current international trade is a global water saving (De Fraiture et al., 2004; Oki and Kanae, 2004; Chapagain et al., 2006a; Yang et al., 2006).

Since the International Expert Meeting on Virtual Water Trade, held in Delft, the Netherlands, in December 2002 (Hoekstra, 2003) and the special session on Virtual Water Trade and Geopolitics during the Third World Water Forum in Japan in March 2003, interest in the concepts of virtual water, water footprints, and global water saving has greatly increased. As a follow-up to the Water Forum in Japan, the World Water Council organized an e-conference on virtual-water trade and geopolitics (WWC, 2004). During three months in Fall 2003 about 300 people participated in a web-based debate about questions such as:

- Does virtual-water trade contribute to the improvement of water availability and through that to local food security, livelihoods, environment, and local economy?
- In which conditions should virtual-water trade be encouraged?
- Does virtual water contribute to conflict resolution or will it increase tensions and conflict potential for those countries relying on trade?
- What governance structures would be necessary to enable a fair virtual-water trade?
- How can the concepts of virtual water and water footprints help in creating awareness about water consumption and saving water by modification of diet?
- What is required from whom to progress on the appropriate and fair introduction and use of the virtual-water concept?

After the e-conference, different workshops addressing virtual-water trade and water footprints were organized: by Stanford University (November 2004, March 2005), the German Development Institute (Bonn, September 2005), the Fourth World Water Forum (Mexico City, March 2006), the Global Water System Project (Bonn, June 2006), and the Institute for Social-Ecological Research (Frankfurt,

July 2006), amongst others. In addition, a considerable number of Master and PhD students from all parts of the world have started to devote their studies to the subject. Apparently the issue of "trade and water" has quite suddenly been recognized as a relevant policy concern and an interesting research field. This book aims to summarize our current knowledge in the field, a somewhat tricky effort given the continuous stream of new research results.

In Chapter 2 we discuss how one can assess the virtual-water content of a commodity and show the resulting estimates for various products. In Chapter 3 we explain how international virtual-water transfers can be quantified and draw the virtual-water balance for each country of the world. Chapter 4 shows how international trade can result in both national and global water savings, but can also occasionally result in global water losses. In Chapter 5 we describe how to assess the water footprint of a nation and show the resulting estimates for all nations of the world. Chapter 6 contains a case study for the Netherlands, a humid country, and Morocco, an arid–semi-arid country. In Chapter 7 we elaborate on the virtual-water transfers within China, which surprisingly go from the water-scarce north to the water-rich south. In Chapter 8 we show the water footprint of coffee and tea consumption, and in Chapter 9 we do a similar but more detailed exercise for cotton. Chapter 10 shows how international trade has made many countries heavily "water dependent" and how water has thus become a geopolitical resource. In the final chapter we explore what sorts of global institutional arrangements are needed to make sure that international trade contributes not only to efficiency in water use, but also to sustainable and equitable water use across the globe.

Chapter 2

How Much Water is Used for Producing our Goods and Services?

In 2003 Novib, the Dutch branch of Oxfam International, asked us to estimate the total volume of water that is used to produce one cup of coffee, and similarly for a cup of tea. The question was to trace the origins of the coffee and tea and estimate the volumes of water used in the various production phases. Although we did indeed look at the water uses during the whole production chain, it was clear from the beginning that the agricultural stage, the stage in which the coffee and tea plants grow and deliver their products, would turn out to be the most water consuming. We had to look at the various origins of coffee and tea, because water requirements in one place are different from those in another place. Focusing on coffee, we also had to translate water use per hectare at field level into water use per ton of coffee cherries yielded, and further into water use per ton of roasted coffee, and finally into water use per cup of coffee. In other words, we had to find out, per coffee-producing country, how many cups of coffee can be annually derived from one hectare and relate that to the water use per hectare. We found out that on average a cup of coffee requires 140 liters of water, mostly rainwater for growth of the coffee plant. Wherever we presented our finding, people had difficulty believing it. Of course we started to let friends, colleagues, and students guess first. It seldom occurred that somebody's guess was even in the right order of magnitude. To some extent we have become famous for our magic number of 140 liters, because this is a nice little fact that people like to remember, to bring up at coffee conversation at work or wherever. Although most people think that we are clever for having calculated this number, some colleagues have also criticized us for misleading people by presenting a meaningless number. The essence of the critique was that even if we were right, what would it matter? Most of the water used for coffee, and also for tea, is rainwater, which comes for free, doesn't it?

Later on we will argue that although water generally does indeed come for free, this does not mean it does not have a value. The amount of rainwater above land varies over time and from place to place, but the global annual volume is more or less constant and sets an upper level to annual freshwater use worldwide. As soon as water is scarce, it has a value (by economic definition). Also, rainwater has a value if there are competing uses that cannot be fulfilled simultaneously. An economist would say: when rainwater is applied for the production of one crop, the opportunity cost of the rainwater is equal to the value that it would have generated if it had been applied for the production of another, more valuable crop. At this point, however, we are not going to follow this line of argument. Instead, we will consider the water need for a few other commodities, to give some factual basis.

Virtual Water

But first some discussion about terminology. We are not unique in asking the question: How much of an input was necessary to produce something? Energy experts for example speak about the "embodied energy" of a good or service, referring to the amount of direct and indirect energy required to produce it (Herendeen, 2004). Similarly one could speak about the amount of land or labor required to make a product, which would then be called embodied land, embodied labor, etc. We could thus speak here about the volume of "embodied water" in a product. Tony Allan originally used the term "embedded water," but since the concept was not well understood he later switched to "virtual water" (Allan, 2003). The term "virtual water" is often loosely defined as the water needed to produce a product. This definition refers to the water *needed*, which is not necessarily the same as the water *actually used*. The "water need" can be more or less than the actual water volume used, depending on the interpretation of what is needed. The concept of water *need* is hypothetical in nature and thus difficult to handle. We prefer to refer to the actual water volume used, which creates the possibility for empirical measurement and validation. In our own studies we apply the following definition: the "virtual-water content" of a product is the volume of water used to produce it, measured at the place(s) where it was actually produced. Some scholars, such as Taikan Oki from Japan, have argued that the virtual-water content of a product should be calculated not at the place of actual production but at the place of consumption (Oki and Kanae, 2004). The argument is that it is particularly useful to know how much water would have been required in the importing country if the goods had not been imported but produced domestically instead. In this philosophy, the term virtual water is therefore defined as "the water volume that would have been required if the product had been produced in the country of consumption." Virtual water is here thus again a hypothetical *need*, not an actual volume, and thus difficult to establish empirically.

In our studies we stick to the idea of defining virtual-water content based on actual water use. Especially when we want to establish the water footprint of a nation we are most interested in the actual water use of the products consumed. For that purpose we need to know the virtual-water content of the products as they are in the producing countries. Only in Chapter 4, when discussing water savings in countries that import water in virtual form, will we come back to this issue, because we will then compare the water that would have been required in an importing country if the product had been produced domestically with the water that was actually used in the exporting country. In the remainder of the book we will follow the principle of measuring the virtual-water content of products at the place of production. The adjective "virtual" refers to the fact that most of the water used in the production is ultimately not contained within the product. The real-water content of products is generally negligible compared with the virtual-water content. The (global average) virtual-water content of wheat for instance is 1,300 m^3/ton, while the real-water content is less than 1 m^3/ton.

Although successful as a concept – most practitioners in the field of water scarcity management nowadays are familiar with it – the term "virtual water" has resulted in considerable confusion and debate. Scholars such as Stephen Merrett who do not approve of the concept like to joke about it as being virtual in itself (Merrett, 2003).

Although we ourselves have contributed to the popularization of the concept, we acknowledge that the term, although apparently appealing to people's imagination, is not very self-explanatory. Alternative terms that have been proposed are, however, not much better. The advantage of a term such as "embodied" or "embedded water" would be that it links up with the term "embodied energy" used in another branch of environmental studies. However, speaking about "embodied water" in a product is confusing in the sense that most of the water used to make a product will not literally be incorporated in the product. In fact the term "embodied water" suggests that it refers to the real-water content of a product, which as we have explained is quite different from its virtual-water content. Munther Haddadin, the former Minister of Water and Irrigation of Jordan, has used the term "exogenous water" as an alternative to "virtual water," referring to the fact that importing water-intensive products can be regarded as an external source of water for the importing country (Haddadin, 2003). He has also used the term "shadow water" (Haddadin, 2006). Both terms have been inspired by the context in which they are used: the discourse on domestic water saving through import of water-intensive commodities. The terms are less useful in a more generic context where we simply want to refer to the volume of water used to produce a product. As it is, the most decisive argument for choosing the term "virtual water" in this book was that, if we had not done so, we would have deviated from what has become common in the world of water experts.

How to Estimate the Virtual-Water Content of an Agricultural Product

The virtual-water content (m^3/ton) of primary crops can be calculated as the crop water use at field level (m^3/ha) divided by the crop yield (ton/ha). The crop water use depends on the crop water requirement on the one hand and the actual soil water available on the other. Soil water is replenished either naturally through rainwater or artificially through irrigation water. The crop water requirement is the total water needed for evapotranspiration under ideal growth conditions, measured from planting to harvest. "Ideal conditions" means that adequate soil water is maintained by rainfall and/or irrigation so that it does not limit plant growth and crop yield. The crop water requirements of a certain crop under particular climatic circumstances can be estimated with the models developed by the Food and Agriculture Organization of the United Nations (Allen et al., 1998; FAO, 2006b). Actual water use by the crop is equal to the crop water requirement if rainwater is sufficient or if shortages are compensated through irrigation. In the case of rainwater deficiency and the absence of irrigation, actual crop water use is equal to effective rainfall. In most of our studies we have assumed that actual water use equals the crop water requirement, since irrigation data specified per crop per country are not easily available. In the cotton study discussed in Chapter 9, however, we have carefully looked per country whether the water requirements of the cotton plant were actually met or not. In some cases we found that actual water use was indeed lower than the crop water requirement.

For calculating crop water requirements one needs various climate and crop data. In the calculations made for this book we have taken country-average data for actual vapor pressure, daily maximum temperature, daily minimum temperature, and percentage cloud cover from the on-line database of the Tyndall Centre for Climate Change and Research (Mitchell, 2004). This database contains averages over the period 1961–90. Data on average elevation, latitude, and wind speed have been taken from the CLIMWAT database of the Food and Agriculture Organization (FAO, 2006c). Crop coefficients for different crops and crop calendars have also been taken from the FAO (Allen et al., 1998; FAO, 2006a). Data on average crop yield (ton/ha) and annual production (ton/yr) per primary crop per country have been taken from the on-line FAOSTAT database (FAO, 2006e).

When a primary crop is processed into a crop product (e.g. paddy rice processed into brown rice), there is often a loss in weight, because only part of the primary product is used. In such a case we calculate the virtual-water content of the processed product by dividing the virtual-water content of the primary product by the so-called product fraction. The product fraction denotes the weight of crop product in tons obtained per ton of primary crop. If a primary crop is processed into two or more different products (for example soybean processed into soybean flour and soybean oil), we need to distribute the virtual-water content of the primary crop amongst its products. We do this proportionally to the value of the crop products. When during processing there is some water use involved, the process water use is added to the virtual-water content of the root product (the primary crop) before the total is distributed over the various root products. In a similar way we can calculate the virtual-water content for products that result from a second or third processing step. The first step is always to obtain the virtual-water content of the input (root) product and the water used to process it. The total of these two elements is then distributed over the various output products, based on their product fraction and value fraction (for a detailed description see Appendix I).

The virtual-water content of live animals can be calculated based on the virtual-water content of their feed and the volumes of drinking and service water consumed during their lifetime (Chapagain and Hoekstra, 2003). As input data one needs to know the age of the animal when slaughtered and the diet of the animal during its various life stages. In our studies we have included eight major animal categories: beef cattle, dairy cows, swine, sheep, goats, fowls/poultry (meat purpose), laying hens, and horses. The calculation of the virtual-water content of livestock products has again been based on product fractions and value fractions, following the above-described method.

Product fractions have been taken from the commodity trees available through the Food and Agriculture Organization (FAO, 2003b). Value fractions have been calculated based on the market prices of the various products. The global average market prices of the different products for the period 1997–2001 have been calculated using trade data from the International Trade Centre (ITC, 2004).

Water Use for Crop and Livestock Products

In general, one finds that livestock products have a higher virtual-water content than crop products. This is because a live animal consumes a large amount of feed crops, drinking water, and service water in its lifetime before it produces some output. Let us consider for example beef produced in an industrial farming system. It takes three years on average before an animal is slaughtered to produce about 200 kg of boneless beef. In this time the animal consumes nearly 1,300 kg of grain (wheat, oats, barley, corn, dry peas, soybean meal, and other small grains), 7,200 kg of roughage (pasture, dry hay, silage, and other roughage), 24 cubic meters of water for drinking, and 7 cubic meters of water for servicing. This means that to produce 1 kg of boneless beef, we use about 6.5 kg of grain, 36 kg of roughage, and 155 liters of water (only for drinking and servicing). Producing the volume of feed requires about 15,340 liters of water on average. The higher we go up in the product chain, the greater will be the virtual-water content of the product. For example, the global average virtual-water content of maize, wheat, and brown rice is 900, 1,300, and 3,000 m³/ton, respectively, whereas the virtual-water content of chicken meat, pork, and beef is 3,900, 4,900, and 15,500 m³/ton, respectively.

Table 2.1 presents estimates for the virtual-water content of some selected crop and livestock products for a number of countries. It can be seen that the virtual-water content of products varies greatly from place to place, depending on the climate and on technology adopted for farming and corresponding yields. The global average virtual-water content of white rice for consumption is about 3,400 m³/ton, but it is much more if the rice comes from India or Brazil. That is: rice from India and Brazil consumes more water *on average*, but within these countries there are large regional differences. One can also see that the virtual-water content of white rice is always larger than the virtual-water content of paddy rice as harvested from the field. The reason is the weight loss that takes place when paddy rice is processed into white rice.

So far we have expressed the virtual-water content of various products in terms of cubic meters of water per ton of product. A consumer might be more interested to know how much water it takes per unit of consumption (Table 2.2). For example, one hamburger requires 2,400 liters of water, the major part of which is used for the beef. One glass of milk requires 200 liters, one glass of wine 120 liters, and one glass of beer 75 liters. Drinking beer preserves water indeed. Even more water-wise however would be to drink plain water!

At the Third World Water Forum in Japan in 2003, we drafted a menu with prices in water units instead of dollars, to make people aware of the hidden link between consumer goods and water use. Obviously, the vegetarian menu requires less water than the non-vegetarian menu. We can observe here a parallel with the required

Table 2.1 Average virtual-water content (m^3/ton) of various products for selected countries.

Product	USA	China	India	Russia	Indonesia	Australia	Brazil	Japan	Mexico	Italy	Netherlands	World average
Rice (paddy)	1,275	1,321	2,850	2,401	2,150	1,022	3,082	1,221	2,182	1,679	—	2,300
Rice (brown)	1,656	1,716	3,702	3,118	2,793	1,327	4,003	1,586	2,834	2,180	—	3,000
Rice (white)	1,903	1,972	4,254	3,584	3,209	1,525	4,600	1,822	3,257	2,506	—	3,400
Wheat	849	690	1,654	2,375	—	1,588	1,616	734	1,066	2,421	619	1,300
Maize	489	801	1,937	1,397	1,285	744	1,180	1,493	1,744	530	408	900
Soybeans	1,869	2,617	4,124	3,933	2,030	2,106	1,076	2,326	3,177	1,506	—	1,800
Sugar cane	103	117	159	—	164	141	155	120	171	—	—	175
Seed cotton	2,535	1,419	8,264	—	4,453	1,887	2,777	—	2,127	—	—	3,600
Cotton lint	5,733	3,210	18,694	—	10,072	4,268	6,281	—	4,812	—	—	8,200
Barley	702	848	1,966	2,359	—	1,425	1,373	697	2,120	1,822	718	1,400
Sorghum	782	863	4,053	2,382	—	1,081	1,609	—	1,212	582	—	2,850
Coconuts	—	749	2,255	—	2,071	—	1,590	—	1,954	—	—	2,550
Millet	2,143	1,863	3,269	2,892	—	1,951	—	3,100	4,534	—	—	4,600
Coffee (green)	4,864	6,290	12,180	—	17,665	—	13,972	—	28,119	—	—	17,000
Coffee (roasted)	5,790	7,488	14,500	—	21,030	—	16,633	—	33,475	—	—	21,000
Tea (made)	—	11,110	7,002	3,002	9,474	—	6,592	4,940	—	—	—	9,200
Beef	13,193	12,560	16,482	21,028	14,818	17,112	16,961	11,019	37,762	21,167	11,681	15,500
Pork	3,946	2,211	4,397	6,947	3,938	5,909	4,818	4,962	6,559	6,377	3,790	4,850
Goat meat	3,082	3,994	5,187	5,290	4,543	3,839	4,175	2,560	10,252	4,180	2,791	4,000
Sheep meat	5,977	5,202	6,692	7,621	5,956	6,947	6,267	3,571	16,878	7,572	5,298	6,100
Chicken meat	2,389	3,652	7,736	5,763	5,549	2,914	3,913	2,977	5,013	2,198	2,222	3,900
Eggs	1,510	3,550	7,531	4,919	5,400	1,844	3,337	1,884	4,277	1,389	1,404	3,300
Milk	695	1,000	1,369	1,345	1,143	915	1,001	812	2,382	861	641	1,000
Milk powder	3,234	4,648	6,368	6,253	5,317	4,255	4,654	3,774	11,077	4,005	2,982	4,600
Cheese	3,457	4,963	6,793	6,671	5,675	4,544	4,969	4,032	11,805	4,278	3,190	4,900
Leather (bovine)	14,190	13,513	17,710	22,575	15,929	18,384	18,222	11,864	40,482	22,724	12,572	16,600

Table 2.2 Global average virtual-water content of selected products, per unit of product.

Product	Virtual-water content (liters)
1 sheet of A4 paper $(80 \, g/m^2)$	10
1 tomato (70 g)	13
1 potato (100 g)	25
1 cup of tea (250 ml)	35
1 slice of bread (30 g)	40
1 orange (100 g)	50
1 apple (100 g)	70
1 glass of beer (250 ml)	75
1 slice of bread (30 g) with cheese (10 g)	90
1 glass of wine (125 ml)	120
1 egg (40 g)	135
1 cup of coffee (125 ml)	140
1 glass of orange juice (200 ml)	170
1 bag of potato crisps (200 g)	185
1 glass of apple juice (200 ml)	190
1 glass of milk (200 ml)	200
1 hamburger (150 g)	2,400
1 pair of shoes (bovine leather)	8,000

land and energy inputs: as we know from other studies, meat also requires more land and energy than a caloric-equivalent amount of crops.

Water Use for Industrial Products

The virtual-water content of an industrial product can be calculated in a similar way as described earlier for agricultural products. There are however numerous categories of industrial products with a diverse range of production methods, and detailed standardized national statistics related to the production and consumption of industrial products are hard to find. As the global volume of water used in the industrial sector is only 716 billion m^3/yr ($\approx 10\%$ of total global water use), we have – for each country – calculated an average virtual-water content per dollar added value in the industrial sector

(m^3/ US$) simply as the ratio of the industrial water withdrawal in a country (m^3/yr) to the total added value of the industrial sector (US$/yr), which is a component of the Gross Domestic Product.

The global average virtual-water content of industrial products is 80 liters per US$. In the USA, industrial products take nearly 100 liters per US$. In Germany and the Netherlands, the average virtual-water content of industrial products is about 50 liters per US$. Industrial products from Japan, Australia, and Canada take only 10–15 liters per US$. In the world's largest developing nations, China and India, the average virtual-water content of industrial products is 20–25 liters per US$.

Water for Domestic Services

Water use for domestic purposes differs greatly from country to country. Roughly speaking, domestic water use per capita increases with increasing gross national income per capita, but the increment becomes less and less at high income levels. At a very low gross national income level of 200 US$ per capita per year (for example Nepal), domestic water use is in the order of 10 m^3 per year per person. At a gross national income level of 1,000 US$ per capita per year (for example Syria), domestic water use is on average 40 m^3 per year per person. At a high-income level of 33,000 US$ per capita per year (for example the USA), domestic water use is as high as 200 m^3 per year per person. There are however many countries that do not conform to the average crude relation just described. The Netherlands for instance is a high-income country, with a gross national income of 25,000 US$ per capita per year, but domestic water use is only 28 m^3 per year per person. Australia on the other hand, also a rich country, has a domestic water use of 340 m^3 per year per person.

Differences between countries can be explained largely by income, climate, technologies applied, and water use habits. The volume of water used in one toilet flush can vary considerably depending on the type of toilet. It can be as low as 6 liters per flush or as high as 30 liters per flush. The presence of a water-saving button can lower the average water use to even less than 6 liters per flush. Taking a bath can consume 100–150 liters. The amount of water used when taking a shower obviously depends on the duration, but also on the type of showerhead used. Depending on the type it takes 5–30 liters per minute. Other typical water-consuming activities are washing of clothes, dishwashing, watering the garden, filling up a swimming pool or pond, and car washing.

Water for domestic services is generally obtained from local sources, simply because artificial transport of water over large distances is very expensive. In a few cases, for example in China, India, and Southern Africa, long-distance inter-basin water transfer projects are being planned or carried out in order to supply cities and industries with sufficient water. But these long-distance supplies are insignificant compared with those related to agricultural products. As we will see in the following chapter, international trade in agricultural products brings huge volumes of water in virtual form from countries such as Australia, Canada, the USA, Argentina, and Brazil to Europe, Japan, Mexico, North Africa, and the Middle East.

Chapter 3

Virtual-Water Flows Between Nations as a Result of Trade in Agricultural and Industrial Products

In the past few years a number of studies have become available showing that the virtual-water flows between nations are substantial. The studies indicate that the global sum of international virtual-water flows must be in the range of 1,000–2,000 billion m³/yr (Hoekstra and Hung, 2002, 2005; Zimmer and Renault, 2003; Chapagain and Hoekstra, 2004; De Fraiture et al., 2004; Oki and Kanae, 2004). This means that 1–2% of the global precipitation above land is used to make commodities for export to other countries. This may sound like a modest fraction, but we talk about a volume that is equal to or more than the annual runoff of huge rivers such as the Congo, the Orinoco, the Yangtze, or the Ganges-Brahmaputra. The various studies had different scopes in terms of product coverage and period of analysis so that the results are difficult to compare. Hoekstra and Hung (2002, 2005), the first global study, considered trade in 38 primary crops in the period 1995–99. Zimmer and Renault (2003) accounted for 29 primary crops and 20 processed crop products, taking data for 2000. Oki and Kanae (2004) took the same year of analysis and looked at five primary crops (maize, wheat, rice, barley, soybean) and three livestock products (chicken, pork, beef). De Fraiture et al. (2004) analyzed trade in cereals only. None of these studies included trade in industrial products. The most comprehensive study on international virtual-water flows to date is our study for the period

1997–2001, covering trade in crop, livestock, and industrial products (Chapagain and Hoekstra, 2004). In this book we draw on this study. For the estimates of international virtual-water flows reported in this book we considered international trade in 285 crop products (covering 164 primary crops) and 123 livestock products (covering eight animal categories: beef cattle, dairy cows, swine, sheep, goats, fowls/poultry, laying hens, and horses). Trade in industrial products was dealt with all-inclusively as well, but in a more crude way, with the average virtual-water content per dollar of traded industrial product as a key parameter.

Virtual-Water "Trade" or "Transfer"?

Initially, when speaking about virtual-water flows between nations as a result of trade, we talked about "virtual-water trade." When the first author introduced the term "virtual-water trade" (Hoekstra and Hung, 2002) and organized an international expert meeting on this subject (Hoekstra, 2003), he did not expect to receive so much criticism from economists as was the case. Several economists pointed out that the term was misleading, because real things are traded, not virtual things. So why not speak about "food trade" instead of "virtual-water trade"? Unfortunately, at times the debate deviated from the intended one about the relation between trade and water management, often ending up as a discussion about terminology. We realized that we had possibly made a mistake by choosing the term "virtual-water trade," and that in order to prevent confusion we would be better speaking neutrally about "virtual-water transfers" or "virtual-water flows," terms that have a physical rather than an economic connotation. Although since 2003 we have avoided the term "virtual-water trade" ourselves, the phrase pops up everywhere nowadays; the term has gained a life of its own. If at times we use the term "virtual-water trade" again in this book, please forgive us and bear in mind that we do not literally refer to a market with traders in virtual water.

How to Assess International Virtual-Water Flows

Virtual-water flows between nations (m^3/yr) can be calculated by multiplying commodity trade flows (ton/yr) by their associated virtual-water content (m³/ton). Most global trade databases contain import and export data per product per country without specifying the trade partners. If we want to know the precise trade relations we can best use the Commodity Trade Statistics database (COMTRADE) of the United Nations Statistics Division, which covers over 90% of world trade. Data from this database are available through the Personal Computer Trade Analysis System (PC-TAS) that can be acquired through the International Trade Centre in Geneva (ITC, 2004). The ITC is the technical cooperation agency of the United Nations Conference on Trade and Development (UNCTAD) and the World Trade Organization (WTO) for operational, enterprise-oriented aspects of trade development. We have used the PC-TAS version for the period 1997–2001. This version covers trade data from 146 reporting countries disaggregated by product and trade partner country. Since the reporting countries also report trade with countries that are not reporting countries themselves, the database effectively covers trade between 243 countries. However, it does not contain data about trade between non-reporting countries. As mentioned above, we have carried out calculations for 285 crop products and 123 livestock products as contained in the database. Recently, the Food and Agriculture Organization in Rome has launched its new on-line trade database that now specifies the trade partners per product in the same way as COMTRADE or PC-TAS (FAO, 2006e). This new database offers a good alternative to the one that we used, or at least gives a possibility to compare data where doubts about the quality of the data exist.

For assessing international virtual-water flows related to trade in industrial products one could theoretically follow the same approach as described above. One would require data on commodity trade flows and virtual-water content for each separate industrial commodity. This has not been done yet, because the variety of industrial commodities is immense and assessing the virtual-water content for each specific commodity would be too laborious. A simpler approach – and the one we have taken – is to assess virtual-water imports and exports by multiplying monetary data on international trade in industrial products (US$/yr) by country-specific data on the average virtual-water content per dollar of industrial products (m³/US$). The latter has been calculated per country by dividing the water withdrawal in the industrial sector (FAO, 2006a) by the added value in the industrial sector (World Bank, 2004). In this approach, all industrial products are included implicitly. Data on trade in industrial products have been taken from the WTO (2004).

International Virtual-Water Flows

Based on our studies we estimate that the international virtual-water flows during the period 1997–2001 amount to 1,625 billion m³/yr on average. A complete overview of virtual-water imports and exports per country is given in Appendix II. The major share (61%) of the virtual-water flows between countries is related to international trade in crops and crop products. Trade in livestock products contributes 17% and trade in industrial products 22%. The total volume of international virtual-water flows includes virtual-water flows that are related to re-export of imported products. The global volume of virtual-water flows related to export of domestically produced products is 1,200 billion m³/yr (Table 3.1). With a total global water use of 7,450 billion m³/yr (the sum of blue and green water use), this means that *16% of the global water use is used not for domestic consumption but for export.* In the agricultural sector, 15% of the water use is for producing export products; in the industrial sector the figure is 34%.

The major gross water exporters are the USA, Canada, France, Australia, China, Germany, Brazil, the Netherlands, and Argentina. The major gross water importers are the USA, Germany, Japan, Italy, France, the Netherlands, the UK, and China. Table 3.2 presents the virtual-water flows for a number of selected countries. It appears that import of water in virtual form can contribute substantially to the total "water supply" of a country. For instance, the Netherlands imports a net amount of (virtual) water equivalent to its annual net precipitation. Jordan imports a volume of water in virtual form equivalent to *five times* its own annual renewable water resources.

A national virtual-water flow balance can be drafted by subtracting the export volume from the import volume. Map 1 in the color section illustrates which countries are net importers of virtual water and which are net exporters. The largest net exporters are Australia and the Americas (Canada, USA, Argentina, and Brazil). The virtual-water export from Canada is mostly connected to the

Table 3.1 International virtual-water flows and global water use per sector. Period: 1997–2001.

	Gross virtual-water flows			
	Related to trade in agricultural products	Related to trade in industrial products	Related to trade in domestic water	Total
Virtual-water export related to export of domestically produced products (10^9 m³/yr)	957	240	0	1,197
Virtual-water export related to re-export of imported products (10^9 m³/yr)	306	122	0	428
Total virtual-water export (10^9 m³/yr)	1,263	362	0	1,625

	Water use per sector			
	Agricultural sector	Industrial sector	Domestic sector	Total
Global water use (10^9 m³/yr)	6,391	716	344	7,451
Water use in the world used not for domestic consumption but for export (%)	15	34	0	16

export of cereals and livestock products. In the USA the export of virtual water is predominantly related to the export of cereals and soybean. The virtual-water export from Argentina and Brazil is strongly linked to the export of soybeans, which are used elsewhere as animal feed. In Brazil, the export of coffee and tea is also relevant to its total virtual-water export. The regions with the largest net import of virtual water are Europe, Japan, Mexico, North Africa, and the Arabian Peninsula. Within Europe, France is an exceptional case, which can be explained by its export of cereals.

Table 3.2 Virtual-water flows for selected countries. Period: 1997–2001.

Country	Gross virtual-water flows (10^6 m^3/yr)								Net virtual-water import (10^6 m^3/yr)			
	Related to trade in crop products		Related to trade in livestock products		Related to trade in industrial products		Total		Related to trade in crop products	Related to trade in livestock products	Related to trade in industrial products	Total
	Export	Import	Export	Import	Export	Import	Export	Import				
Argentina	46,000	3,100	4,180	811	499	1,730	50,600	5,640	−42,900	−3,370	1,230	−45,000
Australia	46,100	3,860	26,400	745	501	4,400	73,000	9,010	−42,300	−25,600	3,900	−64,000
Bangladesh	771	3,670	652	86	162	415	1,590	4,170	2,900	−566	254	2,590
Brazil	53,700	17,500	11,900	1,910	2,210	3,690	67,800	23,100	−36,200	−10,000	1,480	−44,800
Canada	48,300	16,200	17,400	4,950	29,600	14,300	95,300	35,400	−32,100	−12,500	−15,300	−59,900
China	17,400	36,300	5,640	15,200	49,900	11,600	73,000	63,100	18,800	9,610	−38,300	−9,840
Egypt	1,760	11,400	221	1,470	729	711	2,710	13,600	9,690	1,250	−18	10,900
France	43,400	40,600	13,200	11,800	21,900	19,800	78,500	72,200	−2,830	−1,390	−2,110	−6,340
Germany	27,600	59,800	17,400	16,100	25,400	29,800	70,500	106,000	32,100	−1,370	4,340	35,100
India	32,400	13,900	3,410	343	6,750	2,950	42,600	17,200	−18,500	−3,060	−3,800	−25,300
Indonesia	24,800	26,900	371	1,670	310	1,820	25,400	30,400	2,170	1,300	1,510	4,980
Italy	12,900	47,200	14,900	28,300	10,400	13,500	38,200	89,000	34,200	13,400	3,100	50,700
Japan	954	59,000	955	20,300	4,610	18,900	6,510	98,200	58,100	19,400	14,300	91,700
Jordan	97	4,100	165	462	25	228	287	4,790	4,010	297	203	4,510
Korea Republic	997	24,800	3,930	6,100	2,220	8,340	7,150	39,200	23,800	2,170	6,130	32,100
Mexico	11,800	27,000	5,760	13,400	3,790	9,710	21,300	50,100	15,200	7,660	5,920	28,800
Netherlands	34,500	48,600	15,100	7,850	7,890	12,300	57,600	68,800	14,100	−7,300	4,410	11,200
Pakistan	7,380	8,880	612	98	1,530	579	9,520	9,560	1,500	−514	−947	37
Russia	8,300	30,900	2,500	12,200	36,900	2,900	47,700	46,100	22,600	9,740	−34,000	−1,670
South Africa	6,330	7,750	1,310	1,020	912	1,920	8,550	10,700	1,430	−293	1,010	2,150
Spain	18,300	30,500	8,540	5,970	3,750	8,520	30,500	45,000	12,200	−2,570	4,770	14,400
Thailand	38,400	9,760	2,860	1,760	1,660	3,600	42,900	15,100	−28,700	−1,100	1,940	−27,800
UK	8,770	33,700	3,790	10,200	5,110	20,300	17,700	64,200	25,000	6,380	15,200	46,600
USA	135,000	73,100	35,500	32,900	59,200	69,800	229,000	176,000	−61,500	−2,560	10,600	−53,500

It appears that developed countries generally have a more stable virtual-water balance than developing ones. However, countries that are relatively close to each other in terms of geography and development level can have rather different virtual-water balances. Germany, the Netherlands, and the UK are net importers whereas France is a net exporter. The USA and Canada are net exporters whereas Mexico is a net importer. Although the USA has more than three times as much gross virtual-water export as Australia, the latter is the country with the largest *net* export of virtual water in the world.

Virtual-Water Flows Between World Regions

Gross virtual-water flows between and within 13 world regions are presented in Table 3.3. The table shows inter-regional virtual-water transfers insofar as related to agricultural product trade only, because in the case of industrial product trade we could not trace the precise trade relations. The regions with the largest virtual-water export are North and South America. The largest importers are Western Europe and Central and South Asia. The single most important intercontinental water dependency is Central and South Asia (including China and India), annually importing 80 billion m^3 of virtual water from North America. This is equivalent to one seventh of the annual runoff of the Mississippi. Ironically, the African continent, not known for its water abundance, is a net exporter of water to the other continents, particularly to Europe. This can be seen in Map 2, which shows average virtual-water balances over the period 1997–2001 at the level of the 13 world regions. The green colored regions on the map have net virtual-water export, the red colored ones net import. The map shows the biggest virtual-water flows between the different regions insofar as related to trade in agricultural products. The picture for Central and South Asia is dominated by the high net virtual-water import in Japan. The picture for the Former Soviet Union is dominated by Russia. In Map 2 the Former Soviet Union is pictured in red because it is a net importer of agricultural products; in Map 1 nearly all countries in this region are pictured green because the countries are net exporters of virtual water if industrial product trade is also included.

Table 3.3 Average annual gross virtual-water flows between world regions related to the international trade in agricultural products in the period 1997–2001 (10⁹ m³/yr). The gray-shaded cells show the international virtual-water flows within a region.

Exporter \ Importer	Central Africa	Central America	Central and South Asia	Eastern Europe	Former Soviet Union	Middle East	North Africa	North America	Oceania	South America	Southeast Asia	Southern Africa	Western Europe	Total gross export
Central Africa	0.80	0.07	1.73	1.29	0.03	0.26	0.96	0.90	0.06	0.05	1.19	0.17	16.45	23
Central America	0.08	3.13	3.88	0.65	6.14	0.38	0.75	23.98	0.06	0.58	0.23	0.03	10.67	47
Central and South Asia	1.29	0.81	31.53	1.21	4.08	6.67	3.86	4.44	0.37	0.61	16.90	1.37	9.80	51
Eastern Europe	0.01	0.08	0.69	10.77	4.80	2.65	1.08	0.55	0.08	0.10	0.19	0.03	14.15	24
Former Soviet Union	0.01	0.07	3.06	4.47	16.67	5.38	1.26	0.05	0.00	0.30	0.41	0.00	10.54	26
Middle East	0.24	0.11	2.73	0.84	1.46	8.45	3.43	1.01	0.13	0.17	1.86	0.05	6.91	20
North Africa	0.10	0.24	7.09	6.15	2.11	4.32	5.87	8.37	0.17	2.29	3.49	0.52	63.22	98
North America	0.46	40.65	80.18	1.71	2.43	11.22	11.38	35.10	0.96	11.51	13.72	0.79	25.57	201
Oceania	0.34	1.24	29.32	0.33	0.33	6.22	2.13	11.33	12.63	0.67	14.64	1.11	7.76	75
South America	0.39	3.06	19.82	4.23	4.46	8.92	5.08	19.65	0.37	28.09	4.63	1.93	54.44	127
Southeast Asia	1.96	0.50	35.57	2.43	1.52	7.75	8.00	10.89	2.49	0.93	26.82	2.54	18.14	93
Southern Africa	1.04	0.06	2.12	0.38	0.19	0.53	0.54	1.12	0.05	0.17	2.41	2.59	7.21	16
Western Europe	1.40	2.60	15.45	18.87	10.56	12.28	14.26	9.79	0.91	2.45	2.61	1.82	183.51	93
Total gross import	7	50	202	43	38	67	53	92	6	20	62	10	245	895

Are Consumers Co-Responsible for the Effects of Water Use?

Different views exist on the usefulness of visualizing virtual-water flows. What is the *meaning* of showing the connectedness between consumers and the water use that was necessary to produce the consumer goods? In our view consumers are not free from responsibility for the possible negative effects such as water depletion or pollution that result from the production of the goods that they consume, even if production and effects occur in one country and consumption in another. Not everyone agrees with this point of view. When we submitted a manuscript to a scientific journal some time ago, one of the reviewers returned the following comment:

> Japanese and Dutch consumers bear no responsibility for the decisions of American farmers and resource managers. If they did not purchase crops and livestock products from America, they would find those commodities elsewhere, produce them domestically, or reduce their effective demands. It is misleading to suggest that consumers of one nation are responsible for depleting resources in another via the mechanism of voluntary international trade.

In our view, however, both consumers and producers are connected to and bear (at least partial) responsibility for problems caused during the production stage of consumer goods. By showing virtual-water flows, we visualize the (otherwise hidden) connection between consumption and water use. If it appears that the consumption of a certain good in one country relates to a problem of water depletion or pollution in another country, as we show for instance for European cotton consumers and the desiccation of the Aral Sea (Chapter 9), we have an interesting starting point for an analysis of responsibilities and mechanisms that could possibly mitigate the environmental problem. As we view it, the fact that trade is voluntary does not take away responsibilities from consumers and producers. The proverb says that when a consumer buys stolen goods, the receiver is as bad as the thief. This principle is embedded in legislation in many countries. Using this as an analog, why would consumers of products that were produced in an unsustainable way be any better than the producers?

International virtual-water transfers are substantial and likely to increase with continued global trade liberalization (Ramirez-Vallejo and Rogers, 2004). We have shown that, today, 16% of global water use is not for producing products for domestic consumption but for making products for export. Assuming that, on average, agricultural production for export does not cause significantly more or fewer water-related problems (such as water depletion or pollution) than production for domestic consumption, this means that roughly *one sixth of the water problems in the world can be traced back to production for export*. Consumers do not see the effects of their consumption behavior due to the tele-connection between areas of consumption and areas of production. The benefits are at the consumption side; since water is generally grossly under-priced, the costs remain at the production side.

The Relation Between Trade and Water Scarcity

We have estimated the volumes of the international virtual-water flows in the world, but not *explained* them. It would be wrong to suggest that all countries with net import of water in virtual form have so *because* they intend to save domestic water resources. By importing virtual water they will indeed preserve domestic water resources, but this does not imply that the idea of water saving was necessarily the driving force behind the virtual-water imports. International trade in agricultural commodities depends on many more factors than differences in water scarcity in the trading nations, including differences in availability of land, labor, knowledge, and capital, and differences in economic productivities in various sectors. The existence of domestic subsidies, export subsidies, or import taxes in the trading nations will also influence the trade pattern. As a consequence, international virtual-water transfers usually cannot – or can only partly – be explained on the basis of relative water abundances or shortages (De Fraiture et al., 2004; Wichelns, 2004). However, Yang et al. (2003) demonstrated that below a certain threshold in water availability, an inverse relationship can be identified between a country's cereal import and its per capita renewable water resources.

We have shown in this chapter that the current global trade pattern significantly influences water use in most countries of the world, either by reducing domestic water use or by enhancing it. We therefore recommend that future national and regional water policy studies include an assessment of the effects of trade on water policy. For water-scarce countries, it would also be wise to do the reverse: study the possible implications of national water scarcity on trade. In short, strategic analysis for water policy making should include an analysis of expected or desirable trends in international or inter-regional virtual-water flows.

Chapter 4

Water Saving Through International Trade in Agricultural Products

The most direct positive effect of trade in water-intensive commodities is that it generates water savings in the countries or regions that import those commodities. This effect has been discussed in virtual-water studies since the mid-1990s (Allan, 2001; Hoekstra, 2003). These national water savings are equal to the import volumes multiplied by the volumes of water that would have been required to produce the commodities domestically. However, trade not only generates water savings for importing countries, it also means water "losses" for the exporting countries (in the sense that the water can no longer be used for other purposes in the exporting countries). The global net effect of a commodity trade flow between two nations will depend on the actual water volume used in the exporting country compared with the hypothetical water volume that would have been required to produce the commodity in the importing country. There will be net water saving if the trade is from countries with relatively high water productivity (i.e. commodities have a low virtual-water content) to countries with low water productivity (commodities with a high virtual-water content). There will be a net global water loss if the transfer is in the opposite direction.

Virtual-water transfers between nations are one means of increasing the world's water use efficiency. There are three different levels at which water use efficiency can be increased (Hoekstra and Hung, 2002, 2005). At a local level, that of the water user, water use efficiency can be increased by charging prices based on full marginal cost, encouraging water-saving technology, and creating awareness among the water users of the detrimental impacts of water abstractions. At the catchment or river basin level, water use efficiency can be enhanced by re-allocating water to those purposes with the highest marginal benefits. Finally, at the global level, water use efficiency can be increased if nations use their relative water abundance or scarcity to either encourage or discourage the use of domestic water resources for producing export commodities. Whereas much research effort has been dedicated to studying water use efficiency at the local and river basin levels, few efforts have been made to analyze water use efficiency at the global level. As Allan (1999b) argues, there is a tendency to focus on means to increase productive and allocation efficiency, undervaluing the potential of international trade as a means to address the uneven distribution of water across the globe.

As an aside, it is worth mentioning that the uneven distribution of water in space is not the only problem. Variation of water availability in time is also a nuisance, and the reason people create temporary reservoirs behind dams. As Daniel Renault from the Food and Agriculture Organization has pointed out, here is an alternative: storage of water in virtual form can bridge time in the same way as trade in water in virtual form can bridge space. Unsustainable exploitation of water during dry periods can be avoided if the wet periods are optimally used. Temporary storage of food (in the form of grains, but also in the form of livestock) can be a more efficient and more environmentally friendly way of bridging dry periods than building large dams for temporary water storage (Renault, 2003).

But back to the issue of trade. According to international trade theory, nations can gain from trade if they specialize in the production of goods and services for which they have a comparative advantage, while importing goods and services for which they have a comparative disadvantage (Wichelns, 2001, 2004). The economic efficiency of trade in a water-intensive commodity between two countries should be evaluated based on a comparison of the opportunity costs of producing the commodity in each of the trading nations. Export of a water-intensive commodity is attractive if the opportunity cost of producing the commodity is comparatively low. This is the case when there is a relatively high production potential for the water-intensive commodity due for example to relative abundance of water and/or a relatively high water productivity (yield per unit of water input) in the country. Import of a water-intensive commodity (instead of producing it domestically) is attractive if the opportunity cost of producing the commodity is comparatively high, for example because water is relatively scarce and/or water productivity in the country is low.

An evaluation of international trade in water-intensive products based on the theory of comparative advantage would be very complex, because of the large number of countries and of product types involved. In the end, a major sticking point would probably be the lack of reliable data on production potentials for various products per country. Therefore, instead, we will focus on a comparison of water productivities in countries and put trade in that perspective.

A first estimate of the global water saving resulting from international trade between countries with differences in water productivity was made by Oki and Kanae (2004). They considered international trade in five major crops (rice, wheat, maize, barley, and soybean) and three types of meat (chicken, pork, beef) in the year 2000. They estimated that the countries exporting these commodities used 683 billion m^3 of water for producing them. They further estimated that the importing countries would have required 1,138 billion m^3 of water to produce the same volume of commodities within their own borders. This implies a global saving of 455 billion m^3 of water as a result of food trade. The study had some limitations, however. First, it assumed a constant global average crop water requirement throughout the world: 15 mm/day for rice and 4 mm/day for the other crops. Hence, the climatic factor, which plays a major role in the water requirement of a crop, was completely neglected. Second, the study did not take into account the role of the crop coefficient, which is the major limiting factor determining the evaporation from a crop at different stages of crop growth. Another study on global water saving as a result of trade is that of De Fraiture et al. (2004), who estimate that international cereal trade reduces global water use by 164 billion m^3/yr.

In this chapter we present more recent and improved estimates of national and global water savings and losses. We use 1997–2001 as a reference period and take into account differences in climate, yield, and cropping calendar per crop and per country. We cover the international trade in all major crop and livestock products. As in the studies of Oki and Kanae (2004) and De Fraiture et al. (2004)

we focus on quantifying water savings and losses in *physical* units. The calculated physical savings and losses cannot be interpreted in a straightforward manner in terms of *economic* savings or losses. The calculated water savings and losses can be valued positive or negative in an economic sense depending on the wider context only, because the economic efficiency of international trade in agricultural commodities depends on many more factors than differences in water productivity alone, such as scarcity of land, labor, knowledge, and capital, land and labor productivity, domestic subsidies, export subsidies, import taxes, and import quotas in the trading countries.

Method

The virtual-water content of a product is estimated as explained in Chapter 2. First the virtual-water content (m^3/ton) of the primary crop is calculated based on crop water use and yield in the producing country. The crop water use is assumed to be equal to the crop water requirement, which is calculated using the method developed by the Food and Agriculture Organization (Allen et al., 1998; FAO, 2006b). The virtual-water content (m^3/ton) of live animals has been calculated based on the virtual-water content of their feed and the volumes of drinking and service water consumed during their lifetime. The virtual-water content of processed products is calculated based on product fractions (ton of crop product obtained per ton of primary crop or live animal) and value fractions (the market value of one crop or livestock product divided by the aggregated market value of all products derived from one primary crop or live animal). See Appendix I for a more detailed description of the method.

The national water saving (m^3/yr) of a country as a result of trade in a particular product can be assessed by multiplying the net import volume (ton/yr) by the virtual-water content of the product. The latter is now measured in the country itself, because the national water saving through import is determined by the water volume that would have been required if the imported product had been produced domestically. Obviously, a calculated national water saving can have a negative sign, which means a net water loss instead of a saving.

The global water saving (m^3/yr) as a result of one particular trade flow can be calculated as the product of the traded volume (ton/yr) and the difference between the virtual-water content in the importing and the exporting country (m³/ton). The global saving is thus obtained as the difference between the water productivities of the trading partners. The total global water saving can be obtained by summing the global savings of all international trade flows.

The case of global water saving is illustrated with an example of the import of husked (brown) rice to Mexico from the USA (Fig. 4.1). The case of global water loss is shown with an example of export of broken rice from Thailand to Indonesia (Fig. 4.2). For the computation of the total water saving due to international trade in agricultural products, the calculation has been carried out for 285 crop products and 123 livestock products as reported in the database PC-TAS (ITC, 2004) which covers international trade between 243 countries for 1997–2001.

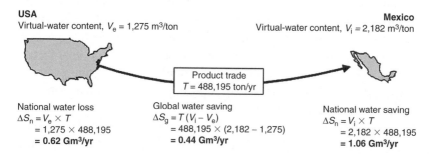

Fig. 4.1 Global water saving through the import of husked rice to Mexico from the USA. Period: 1997–2001.

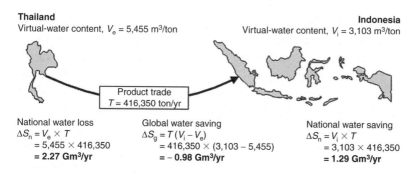

Fig. 4.2 Global water loss through the import of broken rice to Indonesia from Thailand. Period: 1997–2001.

National Water Savings

From our calculations it appears that many countries preserve their national water resources through international trade in agricultural products. Japan annually saves 94 billion m³ from its domestic water resources, Mexico 65 billion m³, Italy 59 billion m³, China (including Hong Kong) 56 billion m³, and Algeria 45 billion m³. The global picture of national savings is presented in Map 3. The data per country are given in Appendix III. The driving force behind international trade in water-intensive products can be water scarcity in the importing countries, but often other factors such as scarcity of fertile land or other resources play a decisive role (Yang et al., 2003). As a result, realized national water savings can only partially be explained through national water scarcity.

The national water saving has different implications in different countries. Though Germany annually saves 34 billion m³, this may be less important from a national policy making perspective because the major products behind the saving are stimulant crops (tea, coffee, and cocoa) which Germany would otherwise not produce. If the import of stimulants was reduced, it would not create any additional pressure on water resources in Germany. However, for Morocco, where import of cereal crop products is the largest national water saver, shifting from import to domestic production would create an additional pressure of 21 billion m³/yr on its national water resources. Table 4.1 shows the nations that save most water through international trade in agricultural products and the main products behind the savings.

For an importing country it is not relevant whether products are consuming green or blue water in the exporting country. The importing country is more interested in what volume and kind of water is being saved from its own resources by the import. It is also important to know whether the water thus saved has higher marginal benefits than the additional cost involved in importing these products.

As an example, Fig. 4.3 shows the national water saving of Egypt as a result of the import of wheat. The mean rainfall of the country is only 18 mm/yr, and almost all agriculture in Egypt is irrigated. At present, Egypt and Sudan base their water resources plans on the division of water agreed by the 1959 Nile water agreement between

Table 4.1 Nations with the largest net water saving as a result of international trade in agricultural products. Period: 1997–2001.

Country	Net national water saving (10^9 m^3/yr)	Major trade partners (10^9 m^3/yr)	Major product categories (10^9 m^3/yr)
Japan	94	USA (48.9), Australia (9.6), Canada (5.4), Brazil (3.8), China (2.6)	Cereal crops (38.7), oil-bearing crops (23.2), livestock (16.1), stimulants (9.2)
Mexico	65	USA (54.0), Canada (5.1)	Livestock (31.0), oil-bearing crops (20.5), cereal crops (19.3)
Italy	59	France (14.6), Germany (6.0), Brazil (5.4), Netherlands (4.4), Argentina (3.1), Spain (3.1)	Livestock (23.2), cereal crops (15.2), oil-bearing crops (12.9), stimulants (8.1)
China	56	USA (17.4), Argentina (8.3), Brazil (8.3), Canada (3.6), Italy (3.4), Australia (3.2), Thailand (2.6)	Livestock (27.5), oil-bearing crops (32.6)
Algeria	45	Canada (10.8), USA (7.6), France (7.1), Germany (4.0), Argentina (1.6)	Cereal crops (33.7), oil-bearing crops (4.0), livestock (3.4)
Russian Federation	41	Kazakhstan (5.2), Germany (4.4), USA (4.1), Ukraine (3.4), Brazil (3.3), Cuba (2.4), France (1.9), Netherlands (1.9)	Livestock (15.2), cereal crops (7.1), sugar (6.9), oil-bearing crops (4.3), stimulants (3.8), fruits (2.3)
Iran	37	Brazil (8.3), Argentina (8.1), Canada (7.7), Australia (6.0), Thailand (2.2), France (2.0)	Cereal crops (22.5), oil-bearing crops (15.1), sugar (1.6)
Germany	34	Brazil (8.3), Côte d'Ivoire (5.3), Netherlands (5.0), USA (4.2), Indonesia (3.3), Argentina (2.2), Colombia (2.1)	Stimulants (21.8), oil-bearing crops (15.0), fruits (3.4), nuts (2.3)
Korea Republic	34	USA (15.6), Australia (3.6), Brazil (2.2), China (1.5), India (1.4), Malaysia (1.2), Argentina (1.1)	Oil-bearing crops (14.3), cereal crops (12.8), livestock (2.3), sugar (1.9), stimulants (1.5)
UK	33	Netherlands (5.3), France (3.7), Brazil (2.8), Ghana (1.9), USA (1.8), Côte d'Ivoire (1.5), Argentina (1.4)	Oil-bearing crops (10.1), stimulants (9.5), livestock (5.2)
Morocco	27	USA (7.8), France (6.4), Argentina (3.3), Canada (2.2), Brazil (1.2), Turkey (0.8), UK (0.8)	Cereal crops (20.9), oil-bearing crops (4.4)

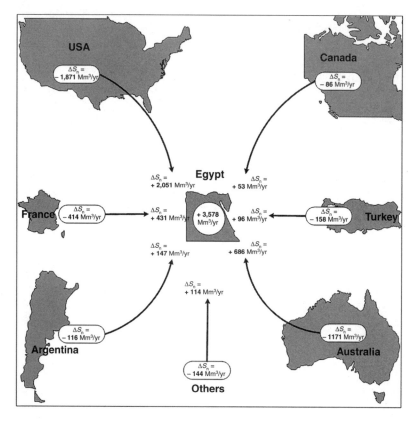

Fig. 4.3 National water saving related to the net import of wheat to Egypt. Period: 1997–2001.

these two countries. However, future developments in upstream countries will have to be taken into account. Disputes over the distribution of water from the Nile could become a potential source of conflict and contention. The expansion of irrigation in the basin will require basin-wide cooperation in the management of water resources to meet increasing demands and to deal with the associated environmental consequences. In this context, the import of wheat in Egypt is contributing to a national water saving of 3.6 billion m^3/yr, which is about 7% of the total volume of water Egypt is entitled to according to the 1959 agreement. The national saving is made with the investment of foreign exchange of 593 million US\$/yr (ITC, 2004). Hence, the cost of the virtual water is 0.16 US\$/m^3 at most. In fact, the cost will be much lower, because the costs of the imported wheat cover not only the cost of water, but also the costs of other input factors such as land, fertilizer, and labor. In Egypt, fertile land is also a major scarce resource. The import of wheat not only releases the pressure on the disputed Nile water, but also reduces pressure to increase the area of land under agriculture. Greenaway et al. (1994) and Wichelns (2001) have shown that in the international context Egypt has a comparative disadvantage in the production of wheat, so that the import of wheat into Egypt implies not only a physical water saving, but also an economic saving.

National Water Losses

Whereas import of agricultural products implies that national water resources are saved, export of agricultural products entails the loss of national water resources. The term "national water loss" is used in this book to refer to the fact that water used for producing commodities that are consumed by people in other countries is no longer available for in-country purposes. The term "water loss" is used here as the opposite of "water saving." As explained earlier, the terms "loss" and "saving" are not to be interpreted in terms of economic loss or saving, but in a physical manner. Water losses as defined here are negative in an economic sense only if the benefit in terms of foreign earning does not outweigh the costs in terms of opportunity cost and negative environmental impacts at the site of production.

The nations with the largest net annual water loss are the USA (92 billion m^3), Australia (57 billion m^3), Argentina (47 billion m^3), Canada (43 billion m^3), Brazil (36 billion m^3), and Thailand (26 billion m^3). Map 4 shows the water losses of all countries that have a net water loss due to production for export. A list of nations with the largest net water loss through international trade in agricultural products is presented in Table 4.2.

The main products behind the national water loss from the USA are oil-bearing crops and cereal crops. These products are grown partly rain-fed and partly irrigated. However, the loss from Côte d'Ivoire and Ghana is mainly from the export of stimulants, which are almost entirely rain-fed. In these countries agricultural use of green water has no major competition with other uses. This type

Table 4.2 Nations with the largest net water loss as a result of international trade in agricultural products. Period: 1997–2001.

Country	Net national water loss (10^9 m^3/yr)	Major trade partners (10^9 m^3/yr)	Major product categories (10^9 m^3/yr)
USA	92	Japan (29.2), Mexico (26.8), China (14.1), Korea Republic (10.1), Taiwan (8.4), Egypt (3.8), Spain (3.7)	Oil-bearing crops (65.2), cereal crops (45.4), livestock (7.8)
Australia	57	Japan (13.7), China (6.0), USA (5.7), Indonesia (4.7), Korea Republic (3.9), Iran (3.3)	Livestock (24.3), cereal crops (23.1), oil-bearing crops (6.8), sugar (4.3)
Argentina	47	Brazil (6.7), China (3.7), Spain (2.4), Netherlands (2.2), Italy (2.1), USA (2.0), Iran (1.9)	Oil-bearing crops (29.9), cereal crops (12.8), livestock (3.7)
Canada	43	USA (12.4), Japan (7.9), China (5.2), Iran (3.7), Mexico (3.4), Algeria (2.1)	Cereal crops (29.3), livestock (12.3), oil-bearing crops (9.6)
Brazil	36	Germany (5.8), USA (5.3), China (4.5), France (4.2), Italy (4.2), Netherlands (3.9), Russian Federation (2.8)	Oil-bearing crops (17.7), stimulants (15.8), livestock (9.3), sugar (9.0)
Côte d'Ivoire	32	Netherlands (5.7), France (4.7), USA (4.5), Germany (4.1), Italy (1.7), Spain (1.5), Algeria (1.4)	Stimulants (32.9), oil-bearing crops (1.5)
Thailand	26	Indonesia (4.7), China (4.4), Iran (2.6), Malaysia (2.5), Japan (2.3), Senegal (1.8), Nigeria (1.7)	Cereal crops (23.6), sugar (5.1), roots and tubers (2.5)
Ghana	17	Netherlands (3.6), UK (3.3), Germany (1.7), Japan (1.6), USA (1.3), France (1.0)	Stimulants (19.1)
India	13	China (2.4), Saudi Arabia (2.0), Korea Republic (1.8), Japan (1.6), France (1.3), Russian Federation (1.3), USA (1.3)	Cereal crops (6.1), stimulants (3.2), livestock (3.0), oil-bearing crops (1.8)
France	9	Italy (6.4), Belgium-Luxembourg (3.8), UK (2.8), Germany (2.1), Greece (1.6), Algeria (1.4), Morocco (1.1)	Cereal crops (21.9), sugar (4.6), livestock (4.2)
Vietnam	8	Indonesia (2.3), Philippines (1.7), Germany (0.4), Ghana (0.4), Senegal (0.4), Singapore (0.4), USA (0.4)	Cereal crops (6.8), stimulants (2.7)

of loss to the national water resources is unlikely to be questionable from an economic perspective, because the opportunity costs of this water are low. The concern is limited to the environmental impacts, which are generally not included in the price of the export products.

The national water losses from France, Vietnam, and Thailand are mainly the result of export of cereal crop products. Thailand for example exports 27.8 billion m^3/yr of water in the form of rice (Fig. 4.4), mostly grown in the central and northern regions (Maclean et al., 2002). The monetary equivalent of rice export is 1,556 million US$/yr (ITC, 2004). Hence, from the loss of its national water, Thailand is generating foreign exchange of 0.06 US$/$m^3$. As much of the rice cultivation in Thailand is done during the rainy season,

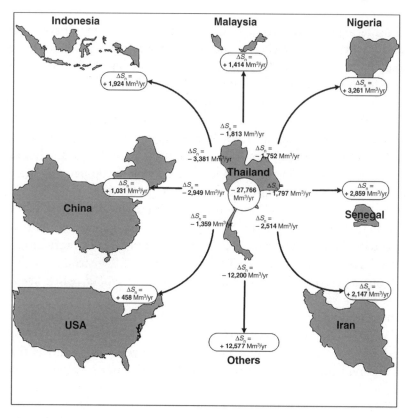

Fig. 4.4 National water loss related to the net export of rice from Thailand. Period: 1997–2001.

the share of green water in the virtual-water content of the rice is quite substantial. Nevertheless, irrigation is widespread, to achieve two harvests per year. If the contribution of irrigation water (blue water) to the total water use of the crop is 50%, and if other resources had zero cost (which is clearly not the case), the value of the blue water used in rice production for export from Thailand would be 0.12 US$/m^3. This represents the upper estimate of the value obtained from the blue water used; since the benefits of rice export should be attributed to all the resources consumed in the production process (not just water, but also land, labor, etc.), the actual value obtained will be much lower.

Global Water Savings

Considering the international trade flows between all major countries of the world and looking at the major agricultural products being traded (285 crop products and 123 livestock products), we have calculated that the actual water use for producing export products amounts to 1,250 billion m^3/yr. If the importing countries were to have produced the imported products domestically they would have required a total of 1,600 billion m^3/yr. This means that the global water saving by trade in agricultural products is 350 billion m^3/yr (Table 4.3). So the average water saving accompanying international trade in agricultural products is (350/1,600 =) 22%. The global volume of water used for agricultural production is 6,400 billion m^3/yr (see Chapter 5). Without trade, supposing that all countries had to produce the products domestically, agricultural water use in the world would amount to 6,750 instead of 6,400 billion m^3/yr. *International trade thus reduces global water use in agriculture by 5%.*

The trade flows that save more than 0.5 billion m^3/yr are shown in Map 5. The trade flows between the USA and Japan and the USA and Mexico are the biggest global water savers. The contribution of different product groups to the total global water saving is shown in Fig. 4.5. Cereal crop products form the largest group, with a saving of 222 billion m^3/yr, followed by oil-bearing crops (68 billion m^3/yr, mainly soybeans), and livestock products (45 billion m^3/yr). The cereal group is composed of wheat (103 billion m^3/yr), maize

Table 4.3 Global water savings as a result of international trade. Period: 1997–2001.

	Related to trade in crop products (10^9 m^3/yr)	Related to trade in livestock products (10^9 m^3/yr)	Total (10^9 m^3/yr)
Global sum of virtual-water exports, assessed on the basis of the virtual-water content of the products in the *exporting* countries	979	275	1,253
Global sum of virtual-water imports, assessed on the basis of the virtual-water content of the products if produced in the *importing* countries	1,286	320	1,605
Global water saving	307	45	352
Actual global volume of water used for agricultural production	—	—	6,391
Global volume of water that would be used for agricultural production if all countries had to produce the products domestically	—	—	6,743
Average water saving accompanying international agricultural product trade flows (%)	—	—	22%
Water saving in global agricultural production as a result of international trade (%)	—	—	5%

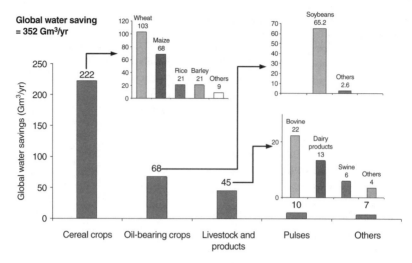

Fig. 4.5 Global water savings (billion m^3/yr) by traded product category. Period: 1997–2001.

(68 billion m^3/yr), rice (21 billion m^3/yr), barley (21 billion m^3/yr), and others (9 billion m^3/yr).

The largest global water savings through wheat trade are occurring as a result of import to the Middle East and North Africa from Western Europe and North America. Map 6 shows the wheat trade flows saving more than 2 billion m^3 of water per year. The global saving of water as a result of maize trade is mainly from the export of maize from the USA. Map 7 shows the maize trade flows saving more than 1 billion m^3/yr. Maize imports in Japan alone are responsible for 15 billion m^3/yr of global water saving. Map 8 shows the global water savings of more than 0.5 billion m^3/yr as a result of rice trade. As the production is more favorable (in terms of climate and culture) in Southeast Asia, the largest savings are from the export from this region to the Middle East and West Africa. The major savings through the trade in rice are the result of trade between Thailand and Iraq, Thailand and Nigeria, Syria and Nigeria, and China and Indonesia.

Considering the import of wheat to Egypt, it can be seen that this contributes to global water saving in some cases and global water loss in others (Fig. 4.3). The import of wheat from the USA, France, and Argentina saves 0.23 billion m^3/yr, whereas the import from

Canada, Turkey, and Australia results in a global water loss of 0.58 billion m³/yr. Though Egypt's import of wheat saves 3.6 billion m³/yr in national water resources, it results in a net global water loss of 0.4 billion m³/yr. The crop water requirement in Egypt is relatively high compared with its trading partners, but this is partly compensated by a relatively high wheat yield, which is more than twice the global average (Table 4.4). As a result, water productivity (production volume per unit of water) in wheat production in Egypt is higher than in Canada, Turkey, and Australia. However, wheat production in Egypt is using scarce blue water resources and the partner countries are making use of the effective rainfall (green water). The net global water loss related to wheat export from Canada, Turkey, and Australia to Egypt results from the fact that the volume of *blue* water resources that would have been required in Egypt to produce domestically is smaller than the volume of *green* water resources actually used in the partner countries. Blue and green water resources fundamentally differ in terms of possible application and thus opportunity cost. As a rule, blue water has higher opportunity costs than green water. For further analysis and interpretation of figures on global water savings and losses it is thus important to separate these figures into a blue and a green water component.

Table 4.4 Crop water requirements, crop yields, and the virtual-water content of wheat in Egypt and its major trade partners. Period: 1997–2001.

Country	Crop water requirement (mm/growing period)	Wheat yield (ton/ha)	Virtual-water content (m³/ton)
Argentina	179	2.4	738
Australia	309	1.9	1,588
Canada	339	2.3	1,491
Egypt	570	6.1	930
France	630	7.0	895
Turkey	319	2.1	1,531
USA	237	2.8	849
Global average	—	2.7	1,334

As a second example let us consider the trade in maize from the USA to Japan. The global water saving from this trade is 15.4 billion m^3/yr. The evaporative demand of maize in Japan (367 mm/growing period) is comparable with that in the USA (411 mm/growing period), but the crop yield in the USA (8.4 ton/ha) is significantly higher than in Japan (2.5 ton/ha), so that the virtual-water content of maize in Japan is three times higher than in the USA. Saving domestic water resources is not the only positive factor for Japan. If Japan wanted to grow the quantity of maize which is now imported from the USA, it would require 6 million hectares of additional cropland – a large amount given the scarcity of land in Japan.

Finally let us consider rice export from Thailand. Though Thailand experiences water losses of 1.7 billion m^3/yr and 1.8 billion m^3/yr by exporting to Nigeria and Senegal, respectively, it is saving water globally as the national water savings in Nigeria (3.2 billion m^3/yr) and Senegal (2.9 billion m^3/yr) are higher than the losses in Thailand (Fig. 4.4). The main reason behind the global saving related to the trade between Thailand and Nigeria is that rice yield in Thailand is 1.7 times higher than in Nigeria (Table 4.5). These two countries have crop water requirements of comparable magnitude (1,000 mm/growing period). By contrast, the main reason behind the global water saving through the trade between Thailand and Senegal, both

Table 4.5 Crop water requirements, crop yields, and the virtual-water content of rice in Thailand and its major trade partners. Period: 1997–2001.

Country	Crop water requirement (mm/growing period)	Rice yield (ton/ha)	Virtual-water content (m^3/ton)
China	830	6.3	1,321
Indonesia	932	4.3	2,150
Iran	1,306	4.1	3,227
Malaysia	890	3.0	2,948
Nigeria	1,047	1.5	7,036
Senegal	1,523	2.5	6,021
Thailand	945	2.5	3,780
USA	863	6.8	1,275
Global average	—	3.9	2,291

of which have a crop yield in the order of 2.5 ton/ha, is the difference in the crop water requirements – 945 mm/growing period in Thailand and 1,523 mm/growing period in Senegal. The export of rice from Thailand to five other trading partners (China, Indonesia, Iran, Malaysia, and the USA) is creating a global water loss of 5 billion m³/yr. National water loss in Thailand is greater than the corresponding national water savings in these countries. This is because rice yield in Thailand is low compared with that in its trading partners.

Global Blue Water Savings at the Cost of Green Water Losses

The total global water saving is made up of a *global blue water saving* and a *global green water saving* component. Even if there is a net global water loss from a trade relation, there might be a saving of blue water at the cost of a greater loss of green water or vice versa. Let us elaborate such a case with the example of Egypt's wheat trade. The virtual-water content of wheat in Egypt is 930 m³/ton. This is all blue water; the green component of the virtual-water content of wheat in this country is zero. Suppose that Egypt is importing T ton/yr of wheat from Australia. The virtual-water content of wheat in Australia is 1,590 m³/ton. Wheat production in Australia is not 100% irrigated; it is assumed here that a fraction f of the virtual-water content of wheat in Australia is green water. The wheat traded from Australia to Egypt causes a global water saving of $(930 - 1,590)T = -660T$ m³/yr. The negative sign means that there is a loss instead of a saving.

The global *green* water saving $S_{g,g}$ (m^3/yr) is always negative as well:

$$S_{g,g} = (0 - 1,590 \times f) \times T = -1,590 \times f \times T$$

Whether the global *blue* water saving $S_{g,b}$ (m³/yr) is positive or negative depends upon the fraction f in the exporting country:

$$S_{g,b} = (930 - 1,590 \times (1 - f)) \times T = (-660 + 1,590 \times f) \times T$$

There is a net saving of global blue water resources as long as the blue water component of Australian wheat is smaller than in Egypt, i.e. if the fraction f in Australia is larger than 0.42. In a case of extreme drought, if the effective rainfall in Australia for wheat is zero ($f = 0$) and all the evaporative demand is met by irrigation, all the losses are in blue water resources, which is $660\,T\,\text{m}^3/\text{yr}$. In another extreme example, when the full evaporative demand of wheat in Australia is met by effective rainfall, so that no irrigation water is used ($f = 1$), the global loss of green water will be $1{,}590\,T\,\text{m}^3/\text{yr}$, but we obtain a net global saving of blue water of $930\,T\,\text{m}^3/\text{yr}$. Here, the gain in blue water is obtained at the cost of green water.

Since blue water resources are generally more scarce than green water resources, global water losses can sometimes still be positively evaluated if blue water resources are being saved. The classic example of trade that makes sense from both water resources and economic points of view is when predominantly rain-fed crop or livestock products from humid areas are imported into a country where effective rainfall is negligible. The import of products that originate from semi-arid countries which apply supplementary irrigation can also be beneficial from a global point of view, because supplementary irrigation increases yields, often to more than double – a profitable situation that could never be achieved in arid countries where effective rainfall is too low to allow for supplementary irrigation, with full irrigation being the only option.

Physical versus Economic Savings

As mentioned above, the volume of the global water saving from international trade in agricultural products is 350 billion m³/yr (average over the period 1997–2001). The largest savings are through trade in crop products, mainly cereals (222 billion m³/yr) and oil-bearing crops (68 billion m³/yr), owing to the large regional differences in virtual-water content of these products and the fact that the products are generally traded from water-efficient to less water-efficient regions. Since there is less variation in the virtual-water content of livestock products, the savings by trade in these products are smaller.

The export of a product from a water-efficient region (relatively low virtual-water content of the product) to a water-inefficient region (relatively high virtual-water content of the product) saves water globally. This is the physical point of view. Whether trade in products from water-efficient to water-inefficient countries is beneficial from an economic point of view depends on a number of additional factors, such as the character of the water saving (blue or green water saving), and the differences in productivity with respect to other relevant input factors such as land and labor. Besides, international trade theory posits that it is not the absolute advantage to a country that indicates what commodities to produce but the comparative advantage (Wichelns, 2004). The decision to produce locally or to import from other sites should be made on the basis of the marginal value or the utility of the water being saved at the consumption site compared with the cost of import.

The Downside of Virtual-Water Import as a Solution to Water Scarcity

Saving domestic water resources in countries with relative water scarcity through virtual-water import (import of water-intensive products) looks very attractive. There are however a number of drawbacks that have to be taken into account. Saving domestic water through import should explicitly be seen in the context of:

- the need to generate sufficient foreign exchange to import food which otherwise would be produced domestically;
- the risk of moving away from food self-sufficiency;
- increased urbanization as import reduces employment in the agricultural sector;
- economic decline and worsening of land management in rural areas;
- reduced access for the poor to food; and
- increased risk of environmental impact in exporting countries, which is generally not accounted for in the price of the imported products.

Increases in virtual-water transfers to optimize the use of global water resources can relieve the pressure on water-scarce countries but may create additional pressure on the countries that produce the water-intensive commodities for export. The potential water saving from global trade is sustainable only if the prices of the export commodities truly reflect the opportunity costs and negative environmental impacts in the exporting countries. Otherwise the importing countries are simply gaining from the fact that they would have had to bear the cost of water depletion if they had produced domestically whereas the costs remain external if they import the water-intensive commodities instead.

Since an estimated 16% of the global water use is not for domestic consumption but for export (see Chapter 3), global water use efficiency becomes an important issue with increasing globalization of trade. Though international trade is seldom done to enhance global water productivity, there is an urgent need to address the increasing global water scarcity problem.

Chapter 5

The Water Footprints of Nations

Reference books and databases on water use traditionally show three columns of water use: water withdrawals in the domestic, agricultural, and industrial sectors, respectively (Gleick, 1993; Shiklomanov, 2000; FAO, 2006a). When you ask a water expert to assess the water demand in a particular country, he or she will generally add together the water withdrawals for the different sectors of the economy. Although useful information, this does not tell us much about the water actually needed by the people in that country in relation to their consumption pattern. The fact is that many goods consumed by the inhabitants of a country are produced in other countries, which means that the real water demand of a population can be much higher than the national water withdrawals suggest. The reverse can also be the case: national water withdrawals might be substantial, but a large proportion of the products are being exported for consumption elsewhere.

In 2002 the first author of this book introduced the concept of the "water footprint" as an indicator of water use that takes the perspective of consumption. The idea was that this indicator could provide useful information in addition to the traditional indicator of water use, which takes the perspective of production. The water footprint of a nation is defined as the total volume of freshwater that is used to produce the goods and services consumed by the people of that nation. Since not all goods consumed in one particular country are produced in that country, the water-footprint consists of two parts: use of domestic water resources and use of water outside the borders of the country. The total water footprint of the inhabitants in a region can be larger than the actual water use within the region, due to net import of water-intensive commodities. National water use in the European Union for example is 559 billion m³/yr from the perspective of production, but 744 billion m³/yr from the perspective of consumption (Table 5.1). On the other hand, annual water use in Australia is 91 billion m³ from the perspective of production, but only 27 billion m³ from the perspective of consumption, which can be explained through the large export trade of water-intensive commodities.

The water footprint has been developed in analogy to the ecological-footprint concept introduced in the 1990s by William Rees and Mathis Wackernagel (Rees, 1992; Wackernagel and Rees, 1996; Wackernagel et al., 1997, 1999). The "ecological footprint" of a population represents the area of productive land and aquatic ecosystems required to produce the resources used, and assimilate the wastes produced, by a certain population at a specified material standard of living, wherever on earth that land may be located. Whereas the "ecological footprint" thus refers to the *area* (hectares) used by an individual or group of people, the "water footprint"

Table 5.1 National water use from the perspectives of production and consumption.

Country/region	National water use* (10^9 m³/year)	
	From the perspective of production[†]	From the perspective of consumption[‡]
Australia	91	27
Canada	123	63
China	893	883
Egypt	59	70
EU25[§]	559	744
India	1,013	987
Japan	54	146
Jordan	1.8	6.3
USA	750	696

* Including both blue and green water use.

[†] Sum of the water withdrawals in the country for domestic and industrial purposes and total crop evapotranspiration within the country.

[‡] The national water footprint of the country as defined in this book.

[§] The 25 countries of the European Union.

indicates the *volume of water* used (cubic meters per year). Most ecological footprint studies are based on global average parameters relating to land requirement per unit of good or service consumed. In our water footprint studies we have from the beginning considered the origin of the goods and services and looked at the actual water use at the place of production.

In this chapter we will start with a brief description of the methodology for assessing the water footprint of a nation (methodological details are provided in Appendix I). The main body of the chapter consists of a review of the actual water footprints of nations of the world.

Two Methods of Assessing the Water Footprint of a Nation

If we know the total water use in a country, we cannot simply assume that this is the water footprint of the country. We have to subtract the part of the water used for making export products that are consumed elsewhere. In other words, we have to subtract the outgoing virtual-water flows. At the same time, however, we have to add the incoming virtual-water flows. This is what we call the top-down approach of assessing the water footprint of a nation. An alternative approach is the bottom-up approach, which multiplies all goods and services consumed by the inhabitants of a country by the respective water needs for those goods and services. So far nobody has ever estimated a national water footprint by applying the bottom-up approach, but there is no reason to assume that it is not possible. In a sense, it is quite straightforward, if data demanding. The national water footprints presented in this book have been calculated through the top-down approach, which was relatively easy because we had already estimated all international virtual-water flows.

The basic input data for our assessment of national water footprints were data on the use of domestic water resources and data on virtual-water inflows and outflows. We have explained in Chapter 3 how we estimated the virtual-water imports and exports per country. With respect to the use of domestic water resources in the industrial and domestic sectors we have drawn on data on water withdrawals in these sectors from the Food and Agriculture Organization (FAO,

2006a). Though significant fractions of domestic and industrial water withdrawals do not evaporate but return to either the ground- or surface water systems, these return flows are generally polluted, so they have been included in the water-footprint calculations. The total volume of water used in the agricultural sector in a country has been calculated based on the total volume of crop produced (FAO, 2006e) and its corresponding virtual-water content (see Chapter 2).

Internal and External Water Footprint

A nation's water footprint has two components, the internal and the external water footprint. The internal water footprint is defined as the volume of domestic water resources used to produce goods and services consumed by inhabitants of the country. It is equal to the total water volume used from domestic water resources in the national economy minus the volume of virtual-water export to other countries insofar as related to export of domestically produced products. The external water footprint of a country is defined as the annual volume of water resources used in other countries to produce goods and services consumed by the inhabitants of the country concerned. It is equal to the so-called virtual-water import into the country minus the volume of virtual water exported to other countries as a result of re-export of imported products.

Both the internal and external water footprints include the consumptive use of *blue water* (originating from ground- and surface water), the consumptive use of *green water* (infiltrated or harvested rainwater), and the production of *gray water* (polluted ground- and surface water). Consumptive use refers to the volume of water that evaporates and thus excludes return flows. Water use in agriculture has been taken as being equal to the evaporation of water at field level during the growing period. The evaporated water originates partly from effective or harvested rainfall (green water) and partly from irrigation water (blue water). We have excluded irrigation losses, assuming that they largely return to the resource base and thus can be reused. We have further neglected gray water production in agriculture. In the industrial and domestic water sectors, the gray water component in the water-footprint calculations has been taken as equal to the polluted return flows.

Water Footprints of Nations

The global water footprint is 7,450 billion m³/yr, which is 1,240 m³/yr per person on average. Humans' green water footprint in the world is 5,330 billion m³/yr, while the blue water footprint amounts to 2,120 billion m³/yr. All the green water goes into agricultural products; the blue water is used for agricultural products (50%), industrial products (34%), and domestic water services (16%).

In absolute terms, India is the country with the largest footprint in the world, with a total footprint of 987 billion m³/yr. However, while India contributes 17% to the global population, the people in India contribute only 13% to the global water footprint. On a relative basis, it is the people of the USA who have the largest water footprint, with 2,480 m³/yr per capita, followed by the people in south European countries such as Greece, Italy, and Spain (2,300–2,400 m³/yr per capita). Large water footprints can also be found in Malaysia and Thailand. At the other end of the scale, the Chinese people have a relatively small water footprint with an average of 700 m³/yr per capita. The average per capita water footprints of nations are shown in Map 9. The data are shown in Table 5.2 for a few selected countries and in Appendix IV for nearly all countries of the world.

The size of the global water footprint is largely determined by the consumption of food and other agricultural products (Fig. 5.1). The total volume of water used globally for crop production is 6,400 billion m³/yr at field level. The estimated contribution of agriculture to the total water use is even greater than suggested by earlier statistics due to the inclusion of green water use (use of rain water). About 83% of the 6,400 billion m³ is green water; the remaining 17% is blue water. The figure of 6,400 billion m³/yr does not include evaporation of irrigation water from storage reservoirs and transport canals. Additional evaporation as a result of human-made surface reservoirs has been estimated to be in the order of 200 billion m³/yr (Shiklomanov, 2000). In addition, about 60% of the annual water withdrawal for agriculture (i.e. 60% of 2,650 billion m³/yr; FAO, 2006a) never reaches the field, because of infiltration or evaporation during transport. If we assume that one fifth of this transport loss consists of evaporation, we have to account for another 300 billion m³/yr of additional evaporation related to agriculture.

Table 5.2 Composition of the water footprint for selected countries. Period: 1997–2001.

Country	Population (10⁶)	Use of domestic water resources (10⁹ m³/yr)						Use of foreign water resources (10⁹ m³/yr)			Water footprint		Water footprint by consumption category (m³/cap/yr)					
		Domestic water withdrawal	Crop evapotranspiration*		Industrial water withdrawal			For national consumption		For re-export of imported products	Total (10⁹ m³/yr)	Per capita (m³/cap/yr)	Domestic water	Agricultural goods		Industrial goods		
			For national consumption	For export	For national consumption	For export		Agricultural goods	Industrial goods				Internal water footprint	Internal water footprint	External water footprint	Internal water footprint	External water footprint	
Australia	19.1	6.51	14.03	68.67	1.229	0.12		0.78	4.02	4.21	26.56	1,393	341	736	41	64	211	
Bangladesh	129.9	2.12	109.98	1.38	0.344	0.08		3.71	0.34	0.13	116.49	896	16	846	29	3	3	
Brazil	169.1	11.76	195.29	61.01	8.666	1.63		14.76	3.11	5.20	233.59	1,381	70	1,155	87	51	18	
Canada	30.7	8.55	30.22	52.34	11.211	20.36		7.74	5.07	22.62	62.80	2,049	279	986	252	366	166	
China	1,257.5	33.32	711.10	21.55	81.531	45.73		49.99	7.45	5.69	883.39	702	26	565	40	65	6	
Egypt	63.4	4.16	45.78	1.55	6.423	0.66		12.49	0.64	0.49	69.50	1,097	66	722	197	101	10	
France	58.8	6.16	47.84	34.63	15.094	12.80		30.40	10.69	31.07	110.19	1,875	105	814	517	257	182	
Germany	82.2	5.45	35.64	18.84	18.771	13.15		49.59	17.50	38.48	126.95	1,545	66	434	604	228	213	
India	1,007.4	38.62	913.70	35.29	19.065	6.04		13.75	2.24	1.24	987.38	980	38	907	14	19	2	
Indonesia	204.9	5.67	236.22	22.62	0.404	0.06		26.09	1.58	2.74	269.96	1,317	28	1,153	127	2	8	
Italy	57.7	7.97	47.82	12.35	10.133	5.60		59.97	8.69	20.29	134.59	2,332	138	829	1,039	176	151	
Japan	126.7	17.20	20.97	0.40	13.702	2.10		77.84	16.38	4.01	146.09	1,153	136	165	614	108	129	

Jordan	4.8	0.21	1.45	0.07	0.035	0.00	4.37	0.21	0.22	6.27	1,303	44	301	908	7	43
Mexico	97.3	13.55	81.48	12.26	2.998	1.13	35.09	7.05	7.94	140.16	1,441	139	837	361	31	72
Netherlands	15.9	0.44	0.50	2.51	2.562	2.20	9.30	6.61	52.84	19.40	1,223	28	31	586	161	417
Pakistan	136.5	2.88	152.75	7.57	1.706	1.28	8.55	0.33	0.67	166.22	1,218	21	1,119	63	12	2
Russia	145.9	14.34	201.26	8.96	13.251	34.83	41.33	0.80	3.94	270.98	1,858	98	1,380	283	91	5
South Africa	42.4	2.43	27.32	6.05	1.123	0.40	7.18	1.42	2.10	39.47	931	57	644	169	26	33
Thailand	60.5	1.83	120.17	38.49	1.239	0.55	8.73	2.49	3.90	134.46	2,223	30	1,987	144	20	41
UK	58.7	2.21	12.79	3.38	6.673	1.46	34.73	16.67	12.83	73.07	1,245	38	218	592	114	284
USA	280.3	60.80	334.24	138.96	170.777	44.72	74.91	55.29	45.62	696.01	2,483	217	1,192	267	609	197
Global total (average)	5,994.3	344	5,434	957	476	240	957	240	427	7,452	1,243	57	907	160	79	40

* Includes both blue and green water use in agriculture.

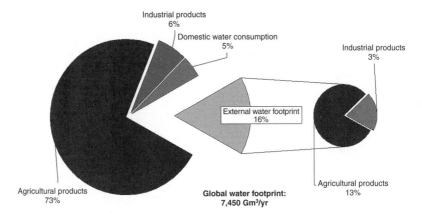

Fig. 5.1 Contribution of different consumption categories to the global water footprint, with a distinction between the internal and the external footprint. Period: 1997–2001.

Rice has the largest share in the total volume of water used for global crop production. It consumes about 1,360 billion m³/yr, which is about 21% of the total volume of water used for crop production at field level. The second largest water consumer is wheat (12%). The contribution of some major crops to total global crop water use is presented in Fig. 5.2. Although the total volume of global rice production is about equal to wheat production, rice consumes much more water per ton of production. The difference is due to the higher evaporative demand in rice production. As a result, the global average virtual-water content of rice (paddy) is 2,300 m³/ton and for wheat 1,300 m³/ton.

Eight countries – India, China, the USA, the Russian Federation, Indonesia, Nigeria, Brazil, and Pakistan – together contribute 50% to the total global water footprint. India (13%), China (12%), and the USA (9%) are the largest consumers of the global water resources (Fig. 5.3).

Both the size of the national water footprint and its composition differ across countries (Fig. 5.4). At one end we see China with a relatively low water footprint per capita, and at the other end the USA. In rich countries consumption of industrial goods makes a relatively large contribution to the total water footprint compared with in developing countries. The water footprints of the USA, China, India, and Japan are presented in more detail in Fig. 5.5. The contribution of the external water footprint to the total water footprint is

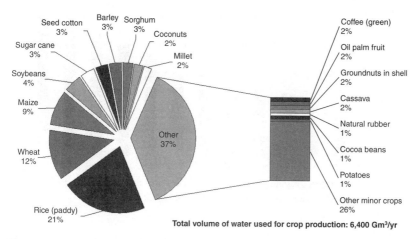

Fig. 5.2 Contribution of different crops to global crop water use. Period: 1997–2001.

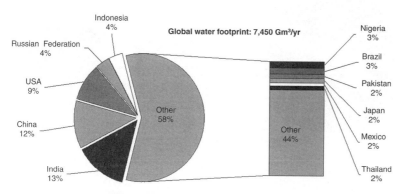

Fig. 5.3 Contribution of major consumers to the global water footprint. Period: 1997–2001.

very large in Japan compared with the other three countries. The consumption of industrial goods contributes significantly to the total water footprint of the USA (32%), but does not contribute much in India (2%).

The share of the external water footprint varies greatly from country to country. Some African countries, such as Sudan, Mali, Nigeria, Ethiopia, Malawi, and Chad, have hardly any external water footprint, simply because they import little. Some European countries on the other hand, for example Italy, Germany, the UK, and

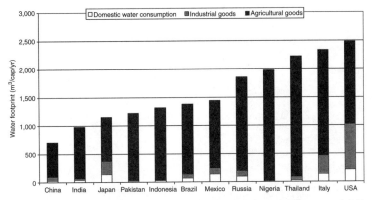

Fig. 5.4 The national water footprint per capita and the contribution of different consumption categories for selected countries. Period: 1997–2001.

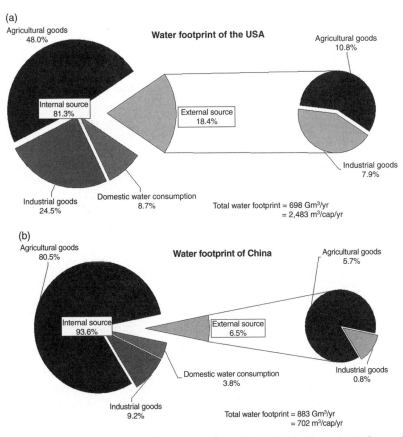

Fig. 5.5 Details of the water footprints of (a) the USA, (b) China, (c) India, and (d) Japan. Period: 1997–2001.

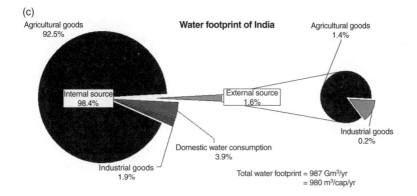

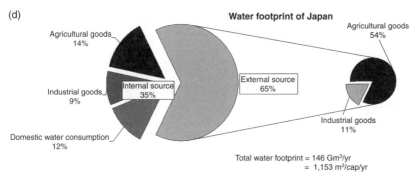

Fig. 5.5 (*Continued*).

the Netherlands, have external water footprints contributing 50–80% to the total water footprint. The aggregated external water footprints of all nations in the world constitute 16% of the total global water footprint (Fig. 5.1). The agricultural products that contribute most to the external water footprints are beef, soybean, wheat, cocoa, rice, cotton, and maize.

Determining Factors

The four major direct factors determining the water footprint of a country are: (i) the volume of consumption (related to the gross national income); (ii) consumption patterns (e.g. high versus low meat consumption); (iii) climate (growth conditions); and (iv) agricultural practice (water use efficiency). In rich countries, people generally consume more goods and services, which immediately translates into larger water footprints. This partly explains the large water footprints of for example the USA, Italy, and Switzerland. But it is not consumption volume alone that determines the water demand of people. The composition of the consumption package is relevant too, because some types of goods require more water than equivalent substitutes. Above all, consumption of meat, in particular beef, contributes to a large water footprint. This factor partly explains the large water footprints of countries such as the USA, Canada, France, Spain, Portugal, Italy, and Greece. Average meat consumption in the USA is 120 kg per year, more than three times the world average. People's water footprint can also be relatively large when the staple food is rice rather than for example wheat or maize. Next to meat and rice consumption, high consumption of industrial goods contributes significantly to the total water footprints of rich countries.

The third factor that influences the water footprint of a nation is climate. In regions with a high evaporative demand the water requirement per unit of crop production is relatively large, leading to greater water footprints in countries such as Senegal, Mali, Sudan, Chad, Nigeria, and Syria. The fourth factor is water use efficiency in agriculture. Inefficient water use increases water use in production, as is evident in countries such as Thailand, Cambodia, Turkmenistan, Sudan, Mali, and Nigeria. In Thailand, for instance, rice yields averaged 2.5 ton/ha in the period 1997–2001, while the global average in the same period was 3.9 ton/ha.

In many poor countries it is a combination of unfavorable climatic conditions (high evaporative demand) and bad agricultural practice (resulting in low water productivity) that contributes to a large water footprint. Underlying factors that contribute to bad agricultural practice and thus large water footprints are the lack of proper water pricing, the presence of subsidies, the use of water-inefficient technology, and lack of awareness among farmers of simple water-saving measures.

The influence of the various determinants varies from country to country. The water footprint of the USA is large (2,480 m³/cap/yr) partly because of high meat consumption per capita and high consumption of industrial products. The water footprint of Iran is relatively high (1,624 m³/cap/yr) partly because of low yields in crop production and partly because of high evapotranspiration. In the USA the industrial component of the water footprint is 806 m³/cap/yr whereas in Iran it is only 24 m³/cap/yr.

How can Water Footprints be Reduced?

Water footprints can be reduced in various ways. One way is to break the seemingly obvious link between economic growth and increased water use, for instance by adopting production techniques that require less water per unit of product. For example, water productivity in agriculture can be improved by applying advanced techniques of rainwater harvesting and supplementary irrigation. A second way of reducing water footprints is to shift to consumption patterns that require less water, for instance by reducing meat consumption. However, it has been debated whether this would be feasible, since the worldwide trend has been for meat consumption to increase rather than decrease. Probably a broader and subtler approach will be needed, where consumption patterns are influenced by pricing, awareness raising, labeling of products, or introduction of other incentives that make people change their consumption behavior. Water costs are generally not well reflected in the price of products due to subsidies in the water sector. Besides, the general public – although often aware of energy requirements – has little awareness of the water requirements in producing their goods and services.

A third method that can be used – not yet broadly recognized as such – is to shift production from areas with low water productivity to areas with high water productivity, thus increasing global water use efficiency. For instance, Jordan has successfully externalized its water footprint by importing wheat and rice products from the USA, which has higher water productivity than Jordan. When importing water-intensive commodities, it would make sense to consider the effect of production in the producing countries. Importing irrigated fruits from Spain, which suffers from severe water shortages, has larger environmental impacts than importing rain-fed fruits from elsewhere. Obtaining irrigated cotton from Uzbekistan or Pakistan is less environmentally friendly than getting rain-fed cotton from the USA. A major problem with this kind of comparison is that we often don't know where a product comes from, let alone about the production conditions of a product. This should change if we want to restore the link between consumers' choices and production circumstances.

The Water Footprint as a New Indicator of Water Use

The water footprint of a nation is an indicator of water use, showing the volumes and locations of water use in relation to the consumption of the people of the nation. It depends on the consumption volume and pattern, but also on the conditions under which the consumer goods are produced. As an aggregated indicator it shows a nation's appropriation of the globe's water resources, a rough measure of the impact of human consumption on the natural water environment. More information about the precise components and characteristics of the total water footprint will be needed, however, before a more balanced assessment of the effects on natural water systems can be made. For instance, one has to look at what is blue versus green water use, because use of blue water often affects the environment more than green water use. It is also relevant to consider the internal versus the external water footprint. For instance, externalizing the water footprint means externalizing the environmental impacts. One also has to realize that some parts of the total water footprint involve the use of water for which no alternative use is possible, while other parts relate to water that could have been used for other purposes with higher value added. There is a difference for instance between beef produced in extensively grazed grasslands of Botswana (use of green water without alternative use) and beef produced in an industrial livestock farm in the Netherlands (fed partly with imported irrigated feed crops).

The figures presented in this chapter largely refer to consumptive water use, i.e. evaporation of rainwater from agricultural fields and evaporation of abstracted surface or groundwater. The effect of water pollution was accounted for to a limited extent by including the (polluted) return flows in the domestic and industrial sector. The calculated water footprints thus consist of three components: green and blue water consumption and wastewater production. The effect of pollution has thus been underestimated, because one cubic meter of wastewater should not count for one, because it generally pollutes many more cubic meters of water after disposal (various authors have suggested factors from 10 to 50). The impact of water pollution can be better assessed by quantifying the water volumes required to dilute waste flows to such an extent that the quality of the water remains above agreed water quality standards. We will illustrate this in Chapter 9, where we estimate the water footprints of nations related to cotton consumption.

Chapter 6

The Water Footprints of Morocco and the Netherlands

Throughout the world freshwater resources have become scarcer over the past few decades, due to increases in population and economic activity and a subsequent increase in water appropriation (Postel et al., 1996; Shiklomanov, 2000; Vörösmarty and Sahagian, 2000; Vörösmarty et al., 2000). In most countries the increase in water use has been largely related to increased production of agricultural products for domestic consumption. However, water use for producing export commodities has also become significant in various countries. As we have shown in Chapter 3, about 16% of the global water use in agriculture in the period 1997–2001 was for producing commodities not for domestic consumption but for export. In certain countries (e.g. Australia, Canada, and Argentina), the agricultural water use for export is actually larger than that for domestic consumption. These countries export water in "virtual" form, i.e. in the form of agricultural commodities. The other side of this phenomenon is that some countries import agricultural commodities instead of producing them domestically, thus importing water in virtual form and saving domestic water resources. Examples are most countries in the Middle East, North Africa, and Europe, but also South Africa, Mexico, and Japan.

In this chapter we will look at two countries with net virtual-water import: Morocco, a semi-arid to arid country, and the Netherlands, a humid country. First, we will consider for both countries the incoming and outgoing virtual-water fluxes. Based on these virtual-water inflows and outflows we assess the water footprints of both countries. Finally, we estimate the water savings (and losses) that result from their international trade. The study is limited to agricultural commodities, since they are responsible for the major part of global water use (see Chapter 5).

Virtual-Water Flows and Balances

In the period 1997–2001 Morocco imported 6.3 billion m³/yr of water in virtual form (in the form of agricultural commodities), and exported 1.6 billion m³/yr. In Morocco itself, water use in the agricultural sector was 37.3 billion m³/yr. The import of cereals was responsible for 3.0 billion m³/yr of virtual-water import. The most important sources of cereals were France, Canada, and the USA. Import of oil crops was the second most important source of virtual-water import into Morocco (1.7 billion m³/yr). Most oil crops were imported from the USA, Argentina, the Ukraine, France, Brazil, and the Netherlands. Other agricultural commodities responsible for significant virtual-water import to Morocco were stimulants (0.7 billion m³/yr) and sugar (0.6 billion m³/yr). An overview of virtual-water flows entering and leaving Morocco is given in Fig. 6.1.

The export of virtual water from Morocco relates particularly to the export of oil crops (0.54 billion m³/yr), fruit (0.32), cereals (0.25), and livestock products (0.23). Italy and Spain are the most important destinations of the oil crops; France and the Russian Federation are the largest customers for fruit; and Libya takes most of the cereals. About 4% of the water used in the Moroccan agricultural sector

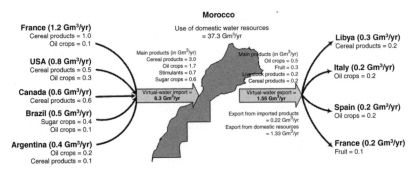

Fig. 6.1 Virtual-water balance of Morocco (insofar as related to trade in agricultural products). Period: 1997–2001.

is applied for producing export products. The remainder of the water is applied for producing products that are consumed by the Moroccan population. From a water security point of view, it seems appropriate that most of the scarcely available water in Morocco is being used for producing commodities for domestic consumption and not for export. From an economic point of view it would be worth checking whether the exported commodities yield a relatively high income of foreign currency per unit of water used. One might doubt for example whether producing cereals for export is a wise thing to do for a water-scarce country like Morocco.

In the period 1997–2001 the Netherlands imported 56.5 billion m^3/yr of water in virtual form (in the form of agricultural commodities) and exported 49.6 billion m^3/yr (Fig. 6.2). Water use in the agricultural sector in the Netherlands itself was 3.0 billion m^3/yr. Imports of stimulants and oil crops were responsible for 18.6 and 18.3 billion m^3/yr of virtual-water import, respectively. The most important sources of stimulants (cocoa, coffee, tea) were Côte d'Ivoire, Ghana, Cameroon, Nigeria, Brazil, Colombia, Kenya, Uganda, and Indonesia. Oil crops came from countries such as the USA, Brazil, and Argentina. Imports of livestock products and cereal products were the third and fourth most import sources of virtual-water import into the Netherlands (7.9 and 6.5 billion m^3/yr, respectively). Most livestock products were imported from neighboring Germany and Belgium. Most cereals

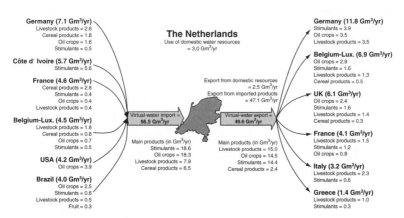

Fig. 6.2 Virtual-water balance of the Netherlands (insofar as related to trade in agricultural products). Period: 1997–2001.

came from France and Germany. Other agricultural commodities responsible for significant virtual-water import to the Netherlands were fruit (1.8 billion m³/yr) and sugar (1.0 billion m³/yr).

Unlike Morocco, the Netherlands has an important through-trade, which means that a large proportion of the imports are exported again in the same or a processed form. As a result, most (about 95%) of the virtual water exported from the Netherlands is not Dutch water, since it can be traced back to countries from which the Netherlands imported. The virtual-water export from the Netherlands related to export of stimulants (not grown in the Netherlands) can for instance be traced back to countries such as Côte d'Ivoire (cocoa) and Brazil (coffee).

Agricultural Water Footprints of Morocco and the Netherlands

Morocco, with a population of 28 million, has an agricultural water footprint of 42.1 billion m³/yr, while the Netherlands, with 16 million inhabitants, has an agricultural water footprint of 9.9 billion m³/yr. Both countries have a significant external water footprint (see Maps 10 and 11). The external water footprint of Morocco is 6.1 billion m³/yr. The water dependency of Morocco – its dependence on foreign water resources, which is defined as the ratio of the external to the total water footprint – is 14%. The water self-sufficiency – the ratio of the internal to the total water footprint – is thus 86%. Morocco mostly depends on virtual-water import from, in sequence, France, the USA, Canada, Brazil, and Argentina.

The total agricultural water footprint of the Netherlands breaks down into an internal footprint of 0.5 billion m³/yr and an external footprint of 9.4 billion m³/yr. The water self-sufficiency of the Netherlands in fulfilling the water needs for the consumption of agricultural commodities is thus 5% and the water dependency 95%. In other words, the total volume of water used outside the Netherlands for producing agricultural products consumed by the Dutch is 20 times the volume of water used in the Netherlands itself. These numbers show the relevance of the water-footprint concept as an alternative indicator of water demand. The agricultural water demand of the Dutch community from a production perspective is

3.0 billion m³/yr (the actual use of water in the agricultural sector in the Netherlands), while the water demand from a consumption perspective is 9.9 billion m³/yr (the global water footprint).

An inhabitant of Morocco has an agricultural water footprint of 1,477 m³/yr on average, while a Dutch person has a footprint of 617 m³/yr. These numbers exclude the volumes of water used for industrial consumer goods and domestic water services (see Appendix IV). As explained in the Chapter 5, the four major factors determining the per capita water footprint of a country are: volume of consumption, consumption pattern, climate, and agricultural practice. The last two factors are unfavorable for the Moroccan water footprint.

Water Savings

Trade between the Netherlands and Morocco generates virtual-water flows from the Netherlands to Morocco and vice versa (Fig. 6.3). The net flow however goes from the Netherlands to Morocco. Morocco uses a small portion of its domestic water resources (50 million m³/yr) for producing fruit, oil crops, nuts, stimulants, and sugar for export to the Netherlands. The flow of virtual water from the Netherlands to Morocco is 140 million m³/yr and is largely related to the trade in cereal products, oil crops, and livestock products. It is worth mentioning here that part of the virtual-water flow from the Netherlands to Morocco does not refer to water use in the Netherlands, because some of the products traded from the Netherlands to Morocco originate from elsewhere. In those cases, the Netherlands is only an intermediate station. For example, the virtual-water flow related to the trade in crude soybean oil (53 million m³/yr) from the Netherlands to Morocco can be traced back to countries such as Brazil and the USA.

If Morocco had to produce domestically the products that are now imported from the Netherlands, this would require 780 million m³/yr of its domestic water resources. Morocco thus saves this volume of water as a result of trade with the Netherlands. The fact that the products imported from the Netherlands were produced with only 140 million m³/yr while they would have required 780 million m³/yr when produced in Morocco means that – from a global perspective – a total water volume of 640 million m³/yr was saved.

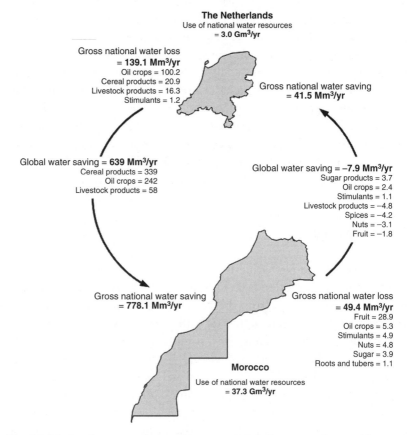

Fig. 6.3 National water savings and losses and global water savings and losses as a result of trade in agricultural products between the Netherlands and Morocco. Period: 1997–2001.

There are two reasons for the large differences in water use per unit of product in Morocco compared with water use per unit of imported product. One is that in the Moroccan climate evaporative demand is relatively high, so that, other circumstances being equal, crops will consume more water than in more moderate climates. The second reason is that current agricultural yields in Morocco are very low (FAO, 2006e). These factors together lead to a situation where maize produced in Morocco has a virtual-water content of 12,600 m³/ton, while maize produced in the Netherlands has a virtual-water content of 410 m³/ton.

If we look at the total virtual-water import of Morocco (6.3 billion m^3/yr, see Fig. 6.1) the domestic water saving is much larger than the domestic water saving related to virtual-water import from the Netherlands alone. According to our calculations, domestically producing the agricultural products that are currently imported to Morocco (period 1997–2001) would require 28.6 billion m^3/yr. Thus, this is the total water volume saved in Morocco as a result of agricultural imports. The global water saving is (28.6 − 6.3 =) 22.3 billion m^3/yr.

Trade in the Context of Managing Water

We have shown above that Morocco and the Netherlands import water in virtual form, more than they export, so that in effect they both partially depend on water resources elsewhere. As visualized with the external water footprints of Morocco and the Netherlands, the consumption of imported products is connected to water use and related impacts in the countries where the products are grown and processed. We also show that the agricultural trade between the Netherlands and Morocco is accompanied by a global water saving. We present these results as analytical facts, without the intention of suggesting that the virtual-water flows revealed are *good* (for example because economically efficient or because saving water resources) or *bad* (because creating dependence or because externalizing negative effects of water use without paying). Although it is tempting to arrive at this sort of judgment, the scope of this book is not broad enough for this kind of conclusion. The evaluation of the question of whether a particular export or import flow is wise or not depends on additional economic factors to the water input and would also include various sorts of political, historical, and cultural considerations. From a water resources point of view alone, however, the overall trade balance of Morocco looks quite positive. The export of water in virtual form is relatively small and primarily in the form of high-value crops. In the future Morocco will face the question of how to find a balance between increasing domestic water productivity and promoting further increase of virtual-water imports. For a semi-arid to arid country like Morocco, an essential political question is: To what extent does it care about food self-sufficiency (producing its own food based on domestic water resources)? Does it care about the use of domestic water resources to produce export products, products which may give high return in terms of foreign currency? Due to the limited availability of water, striving for food self-sufficiency will soon conflict with using water for producing export products. If food self-sufficiency were not an issue, from an economic point of view it would make sense to stimulate export of products with a relatively high foreign currency income per unit of water used (e.g. citrus fruit, olives) and to import products that would otherwise require a relatively large amount of domestic water per dollar produced (e.g. cereals).

The trade balance of the Netherlands is also good from a water resources point of view. The Dutch people would simply not be able to maintain their present consumption pattern if all consumer goods had to be produced domestically. There would be insufficient water, not to mention shortage of land. Besides, some imported products, such as rice, coffee, tea, and cocoa, cannot even be produced in the Netherlands. For the Netherlands the question is rather whether it is fair to depend so heavily on foreign water resources without properly covering the costs that are associated with that foreign water use.

Although we found that both Morocco and the Netherlands have net import of virtual water, it would be unreasonable to say that Morocco and the Netherlands import water in virtual form *because* they purposely intend to save domestic water resources. There might be an indirect relation, but trade patterns are too complex to be explained by a single factor. By importing virtual water both countries indeed save domestic water resources, but this does not imply that the latter was an important incentive for the former. So we fall short in explaining why the two countries have net virtual-water import and in collecting grounds for judging the current trade in terms of positive and negative implications. What we do show, however, is that international agricultural trade can significantly influence domestic water demand and thus domestic water scarcity, and that formulating international agricultural trade policy should therefore include an analysis of the implications in the water sector. The message is: international trade in agricultural products significantly influences the water appropriation in a country, a relation that has so far received little attention from economists and water managers alike.

Chapter 7

Virtual- versus Real-Water Transfers Within China

Mao Zedong had a well-known saying: "There is water abundance in the South and scarcity in the North; if possible we can borrow a little bit water from South to North." Research on the idea of transporting water from South to North China on a large scale started 50 years ago. Today, final plans have been drafted and elements of the South–North Water Transfer Project are being implemented. Indeed, North China is suffering from water shortage and relies on water transfer from the South to relieve the water crisis. However, at the same time, North China, as China's breadbasket, annually exports substantial amounts of water-intensive products to South China. This creates the paradox that huge volumes of water are being transferred from the water-rich South to the water-poor North while large amounts of food are being transferred from the food-sufficient North to the food-deficient South. This paradox is receiving increasing attention, but the research in this field gets stuck at the stage of rough estimation and qualitative description, partly due to the absence of appropriate methodologies to address the issue.

In this chapter we quantitatively assess the virtual-water flows between the regions in China and put them in the context of water availability in each region. We consider China Mainland excluding Hong Kong, Macao, and Taiwan. This area consists of 31 provinces, municipal cities, and autonomous regions. The country is schematized into two main regions: North and South. Both North and South are further divided into four sub-regions. Each sub-region consists of three to five provinces (Table 7.1).

Table 7.1 Regional schematization of China.

Region	Sub-region	Provinces
North China	North-central	Beijing, Tianjin, Shanxi
	Northeast	Inner Mongolia, Liaoning, Jilin, Heilongjiang
	Huang-huai-hai	Hebei, Henan, Shandong, Anhui
	Northwest	Shannxi, Gansu, Qinghai, Ningxia, Xinjiang
South China	Southeast	Shanghai, Zhejiang, Fujian
	Yangtze	Jiangsu, Hubei, Hunan, Jiangxi
	South-central	Guangdong, Guanxi, Hainan
	Southwest	Chongqing, Sichuan, Guizhou, Yunnan, Tibet

Assessing Virtual-Water Flows Between Regions in China

When estimating the virtual-water flows between regions within China we apply the same method used when assessing the virtual-water flows between nations (see Chapter 3). A drawback is that inter-regional trade flows for China are not available, so that they had to be estimated based on regional production surpluses and deficits. Net import of food into a region (or net export from the region) is a function of regional production, domestic utilization, and stock changes. We draw regional food balances using the same definitions as in the food balance sheets of the Food and Agriculture Organization (FAO, 2006e); see Table 7.2. Net import of a product into a region is here taken to be equal to "total domestic supply" of the product (by definition equivalent to "total domestic utilization") minus the production of the product and minus the regional stock change. The net virtual-water import related to trade in a particular product is equal to the net import volume of the product multiplied by its virtual-water content as in the exporting region.

Table 7.2 The structure of a food balance sheet (FAO, 2006e).

Domestic supply					$\longrightarrow$ Domestic utilization						
Production	Import	Stock change	Export	Total	Feed	Seed	Processing	Waste	Other use	Food	Total

Since most of the water use in China, as in most countries, is in agricultural production, we focus here on agricultural products. We have classified these products into six categories: grains, vegetables, fruit, meat (including poultry products), eggs, and dairy products (including milk). The analysis has been carried out with data for the year 1999, when China experienced a normal hydrological year, but a good year in terms of harvest.

In order to proceed we had to make a number of assumptions:

• For each region there is no significant change in product stock at the end of the year.
• Agricultural products imported from outside China go to the provinces with production deficits and are distributed in proportion to the deficit per province.
• Agricultural products exported from China to other nations come from provinces with production surpluses.
• After accounting for international trade, the sub-regions with deficits import agricultural products from the closest neighboring sub-regions with surpluses.

Virtual-Water Content per Product Category per Region

The virtual-water content was first calculated separately for 25 sorts of crops, six kinds of meats, and for eggs and milk. Calculations were carried out per province, based on the methodology explained in Appendix I. Subsequently, the average virtual-water content for each of the six categories of agricultural products was calculated based on a production-weighted average of the virtual-water content of the various products per category. The results are summarized in Table 7.3. In general the virtual-water content of products in North China is higher than in South China. This is a result of the fact that North China, except for part of the Northeast, is arid or semi-arid. The value of the aridity index (the ratio of potential evaporation to precipitation) is larger than 3 or even larger than 7 in some parts of the Northwest. By contrast, South China is humid or semi-humid. Abundant sunshine and strong evaporation in North China, with a similar yield as in South China, mean that the virtual-water content of agricultural products is higher in North China.

Table 7.3 Virtual-water content (m^3/kg) of six agricultural product categories in China by region.

Region	Grains	Vegetables	Fruit	Meat	Eggs	Dairy products
North	1.1	0.1	1.1	7.8	4.3	1.9
South	0.9	0.1	0.8	5.7	4.3	1.9
National average	1.0	0.1	1.0	6.7	4.3	1.9

Food Trade Within China

In the past, South China was the country's breadbasket. According to a Chinese saying: "After harvest both in Hunan and Hubei, the whole country will have sufficiency." However, since the early 1990s this situation has changed. In 1999 North China produced 53% of the nation's grains, 57% of vegetables, 55% of fruit, 48% of meat, 71% of eggs, and 82% of dairy products. Conversely, the consumption in South China of all these agricultural products exceeded 50% of the national total. What was the cause of this shift of breadbasket from South to North China? First, South China, especially the South-central and Southeast, where the reform and open-door policies were initially carried out, is densely populated and the richest area in China. Huge investment has resulted in manufacturing prosperity and the construction of infrastructure that occupies substantial areas of fertile land. Consequently, the nature of the workplace has shifted from agricultural to secondary and tertiary industries. Second, improvements in living standards have caused a dietary change resulting in consumption of more agricultural products. The subsequent huge increase in food demand has stimulated farmers' enthusiasm in areas such as the Northeast and Huang-huai-hai, where the fertile land, sunshine, and heat provide good conditions for improving food production. Also, the national food policy – keeping food self-sufficiency at a high level – determined that a new breadbasket should be established to substitute for the old one in order to feed the country's huge population. These developments have caused the current situation in which virtual water and real water inversely flow between the North and the South of China.

Table 7.4 Food trade in China in 1999 (10^6 ton).

		Grains	Vegetables	Fruit	Meat	Eggs	Dairy products
Net import from other nations	North	−4.2	−2.3	−0.3	−0.2	−0.01	0.00
	South	4.5	−0.1	0.1	0.1	0.00	0.12
	China total*	0.3	−2.4	−0.2	−0.1	−0.01	0.12
Net trade from North to South		17.1	23.2	0.6	1.8	2.3	2.4

* *Source:* ITC (2002, 2004).

At a national level China can currently realize more or less a balance between food demand and production. At the regional level, however, North China has a food surplus and South China a food deficit. As to the eight sub-regions, the food-surplus areas include the Northeast and Huang-huai-hai in North China and Yangtze in South China. The other five regions have food deficits, which in some densely populated and developed areas such as the North-central, Southeast, and South-central accounted for more than 20% of their total demand. In total, South China imported 17 million tons of grains, 23 million tons of vegetables, 0.6 million tons of fruit, 1.8 million tons of meat, 2.3 million tons of eggs, and 2.4 million tons of dairy products from North China (Table 7.4).

Virtual-Water Transfers Within China

Considering trade in agricultural products only, China as a whole had a positive virtual-water balance, with a net import of virtual water in 1999 of 9 billion m^3, equating to around 7 m^3 per person. The gross import was 28 billion m^3, with most virtual water going to the South. The gross export was 19 billion m^3, with the major share originating from the water-scarce North. The virtual-water flow from North to South was around 52 billion m^3 (Table 7.5). The various virtual-water flows between the eight sub-regions are shown in Fig. 7.1. The major exporters of virtual water are the Northeast and Huang-huai-hai; the major importers are the Southeast and South-central.

Table 7.5 Virtual-water imports and exports of China by region in 1999.

	North–South trade (10⁹ m³/yr)	International trade (10⁹ m³/yr)			Overall net virtual-water import (10⁹ m³/yr)	Net virtual-water import per capita (m³/cap/yr)
	Net virtual-water import	Gross virtual-water import	Gross virtual-water export	Net virtual-water import		
North	−51.6	8.4	16.2	−7.8	−59.4	−102
South	51.6	19.7	2.7	17.0	68.6	104
National total	—	28.1	18.9	9.2	9.2	7

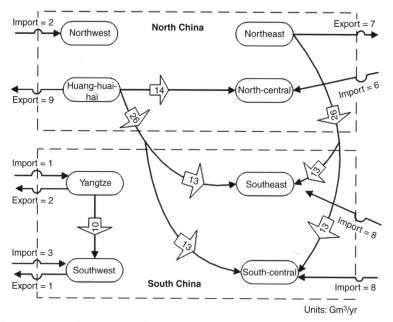

Fig. 7.1 Virtual-water transfers in China in 1999.

Virtual- versus Real-Water Budgets

In 1999, the people of North China transformed 707 billion m³ of their real-water budget of 2,115 billion m³ into virtual water (Fig. 7.2). Together with the import of virtual water from abroad,

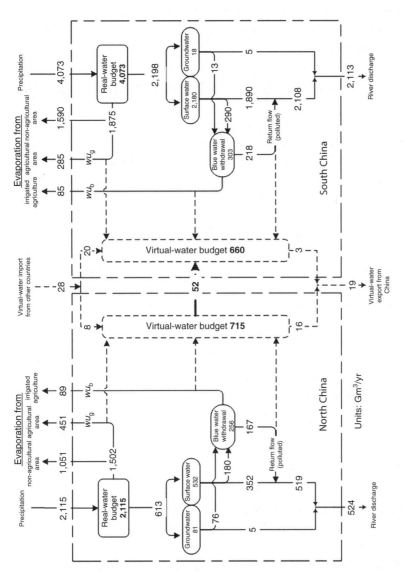

Fig. 7.2   Real- and virtual-water balance of North and South China in 1999.

the virtual-water budget of North China amounted to 715 billion m³. A virtual-water volume of 52 billion m³ was exported to South China and 16 billion m³ to other countries. South China transformed only 588 billion m³ of their real-water budget of 4,073 billion m³ into virtual water. In 1999, the water-scarce North had 100% water self-sufficiency, whereas the water-rich South relied on virtual-water import, having a water self-sufficiency of 90%. In North China, groundwater accounted for 30% of the total blue water withdrawal, which approached 95% of its groundwater flow.

Virtual-Water Transfers in Relation to Water Availability

Studying net virtual-water import per sub-region, one finds that in general the higher the per capita water availability in a sub-region, the larger the volume of virtual-water import (Table 7.6). This is of course very counter-intuitive. Huang-huai-hai, for instance, has a population of 310 million and a water availability of 530 m³ per person per year, which is less even than in the Middle East and North Africa, where 300 million people use the limited water supplies with a per capita share of 900 m³/yr (Berkoff, 2003). Nevertheless, virtual-water export from this region, which is regarded as one of the most water-scarce territories in the world, is quite substantial. Huang-huai-hai exported 49 billion m³ of water in virtual form in 1999, which is 157 m³ per person per year.

As water is of vital importance to agriculture, one would expect – from the perspective of water – that virtual-water export would be proportional to water availability. In China one can find the reverse situation. Apparently factors other than water – perhaps availability of fertile land, but possibly also political factors – have been determinants of the process that has led to the current situation. Even today, the approach is mainly supply oriented. Although concepts of water demand management have long been promoted, they are rarely applied.

North–South Virtual-Water Flows in Relation to the South–North Water Transfer Project

The South–North Water Transfer Project is the biggest inter-basin water transfer project in the world. After a 50-year study, three water-diverting routes have been identified, i.e. the west route, the middle route, and the east route. They will divert water from the upper, middle, and lower reaches of the Yangtze River, respectively, with a maximum transfer amount of 38–43 billion m³/yr (east route 15 billion m³, middle route 13 billion m³, west route 10–15 billion m³) to meet the increasing water requirements of North China (Qian et al.,

Table 7.6 Net virtual-water import in China per region compared with the use of domestic water resources, in 1999.

Sub-region	Use of domestic resources (10⁹ m³/yr)				Net virtual-water import (10⁹ m³/yr)	Net virtual-water import per capita (m³/cap/yr)	Water availability per capita (m³/cap/yr)
	Blue water		Green water	Total			
	Surface water	Groundwater					
North-central	5.5	7.0	26.6	39.1	20.4	375	369
Huang-huai-hai	56.4	33.9	255.4	345.7	−48.7	−157	532
Northeast	49.1	23.0	122.9	195.0	−32.3	−249	1,568
Northwest	68.7	12.4	45.6	126.7	1.2	14	2,487
North total	179.7	76.3	450.5	706.5	−59.4	−102	1,047
Yangtze	118.1	5.9	116.6	240.6	−10.7	−45	1,821
Southeast	48.4	1.1	25.9	75.4	33.5	361	2,976
South-central	74.2	3.7	41.2	119.1	33.8	265	3,143
Southwest	49.4	2.9	101.1	153.4	12.0	61	5,496
South total	290.1	13.6	284.8	588.5	68.6	104	3,347
National total	469.8	89.9	735.3	1,295.0	9.2	7	2,234

2002). The east and middle routes have already been constructed, covering seven provinces and the municipal cities of Beijing, Tianjin, Hebei, Henan, Shandong, Anhui, and Jiangsu.

In 1999 North China exported 52 billion m^3 of virtual water to South China, in the form of agricultural products. Huang-huai-hai, a recipient area of the east and middle routes, had a virtual-water export to South China of 26 billion m^3. Although the maximum transfer volume via the two routes is 28 billion m^3, this also includes the water supply to other provinces in different sub-regions, such as Beijing, Tianjin, and Jiangsu.

The current North–South virtual-water transfer exceeds the planned South–North real-water transfer volume of 38–43 billion m^3/yr. One can look at this in two ways. Either one can conclude, as Chinese officials tend to do, that the planned real-water transfer from the South to the North is not overdesigned and might even be insufficient to meet the water needs in the North. Or one can argue, as some non-governmental organizations do, that reducing the production for export in the North and investing in increased production in the South would be a more straightforward solution than continuing on the path of producing in a region where water is insufficient and implementing massive water transfer projects.

The big question is of course: Is bringing huge volumes of real water from South to North worth its social and environmental consequences? From a water resources perspective it looks odd to transfer water in real form from one place to another and then bring it back in virtual form. Why not try to reduce the use of water in the water-scarce North and gradually reduce the production of water-intensive commodities that are traded to the water-rich South? There must be other decisive factors justifying the current strategy. Factors that could play a role are availability of suitable cropland, labor availability, or national food security. A broader, integrated study is required to give a more comprehensive assessment of the efficiency and sustainability of the South–North Water Transfer Project.

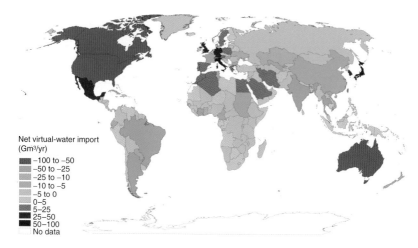

Net virtual-water import
(Gm³/yr)
- −100 to −50
- −50 to −25
- −25 to −10
- −10 to −5
- −5 to 0
- 0−5
- 5−25
- 25−50
- 50−100
- No data

Map 1. Virtual-water balance per country over the period 1997–2001. The balances are drawn based on an analysis of international virtual-water flows associated with trade in both agricultural and industrial products. The red-colored countries have net virtual-water import; the green-colored countries have net virtual-water export. (See p. 22.)

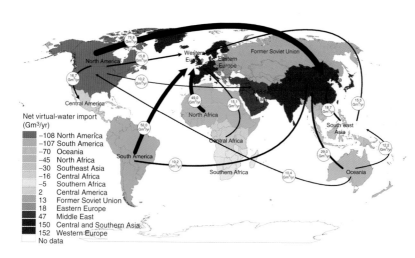

Net virtual-water import
(Gm³/yr)
- −108 North America
- −107 South America
- −70 Oceania
- −45 North Africa
- −30 Southeast Asia
- −16 Central Africa
- −5 Southern Africa
- 2 Central America
- 13 Former Soviet Union
- 18 Eastern Europe
- 47 Middle East
- 150 Central and Southern Asia
- 152 Western Europe
- No data

Map 2. Regional virtual-water balances and net inter-regional virtual-water flows related to trade in agricultural products. Only the largest net flows (> 10 billion m³/yr) are shown. Period: 1997–2001. (See p. 25.)

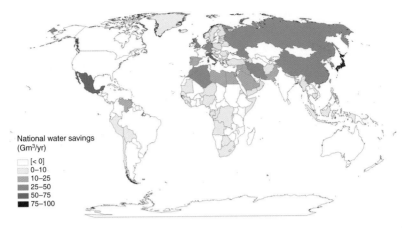

National water savings
(Gm³/yr)

- [< 0]
- 0–10
- 10–25
- 25–50
- 50–75
- 75–100

Map 3. National water savings related to international trade in agricultural products. Period: 1997–2001. (See p. 36.)

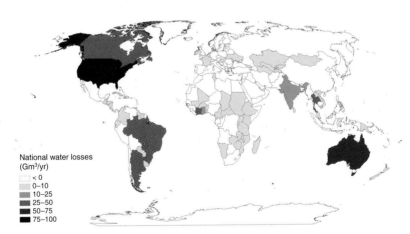

National water losses
(Gm³/yr)

- < 0
- 0–10
- 10–25
- 25–50
- 50–75
- 75–100

Map 4. National water losses related to international trade in agricultural products. Period: 1997–2001. (See p. 39.)

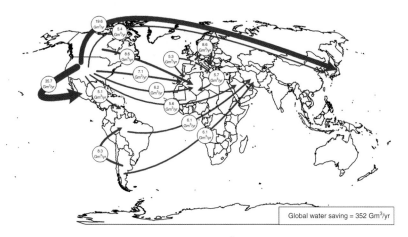

Global water saving = 352 Gm³/yr

Map 5. Global water savings (> 5.0 billion m³/yr) associated with international trade in agricultural products. Period: 1997–2001. (See p. 42.)

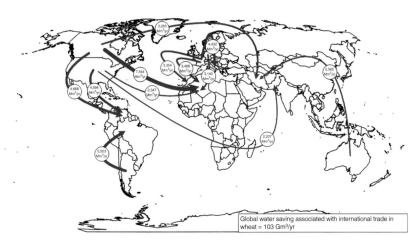

Global water saving associated with international trade in wheat = 103 Gm³/yr

Map 6. Global water savings (> 2.0 billion m³/yr) associated with international trade in wheat. Period: 1997–2001. (See p. 44.)

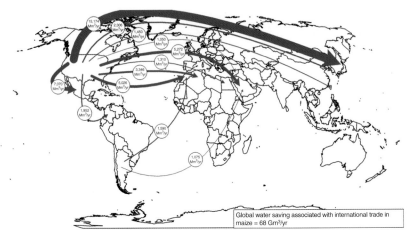

Global water saving associated with international trade in
maize = 68 Gm³/yr

Map 7. Global water savings (> 1.0 billion m³/yr) associated with international trade in maize. Period: 1997–2001. (See p. 44.)

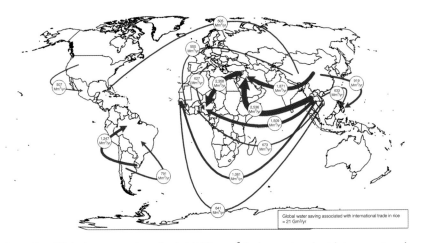

Global water saving associated with international trade in rice
= 21 Gm³/yr

Map 8. Global water savings (> 0.5 billion m³/yr) associated with international trade in rice. Period: 1997–2001. (See p. 44.)

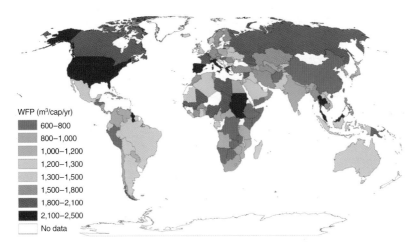

WFP (m³/cap/yr)

- 600–800
- 800–1,000
- 1,000–1,200
- 1,200–1,300
- 1,300–1,500
- 1,500–1,800
- 1,800–2,100
- 2,100–2,500
- No data

Map 9. Average water footprint (WFP) per capita per country. Green-colored countries have a water footprint per capita equal to or less than the global average; red-colored countries are above the global average. Period: 1997–2001. (See p. 55.)

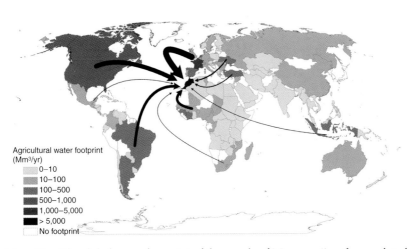

Agricultural water footprint (Mm³/yr)

- 0–10
- 10–100
- 100–500
- 500–1,000
- 1,000–5,000
- > 5,000
- No footprint

Map 10. The global water footprint of the people of Morocco (insofar as related to the consumption of agricultural products). Period: 1997–2001. (See p. 70.)

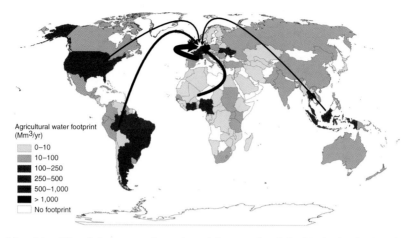

Map 11. The global water footprint of the people of the Netherlands (insofar as related to the consumption of agricultural products). Period: 1997–2001. (See p. 70.)

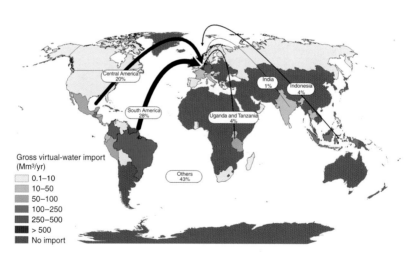

Map 12. Import of water in virtual form into the Netherlands related to coffee imports. Period: 1995–1999. (See p. 95.)

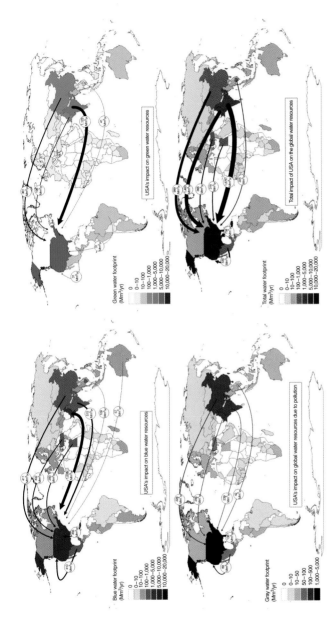

Map 13. The impact of consumption of cotton products by US citizens on the world's water resources (million m³/yr). Period: 1997–2001. (See p. 123.)

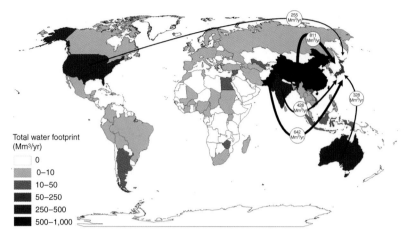

Map 14. The impact of consumption of cotton products by Japanese citizens on the world's water resources (million m³/yr). Period: 1997–2001. (See p. 126.)

Map 15. The impact of consumption of cotton products by the people of the 25 countries of the European Union (EU25) on the world's water resources (million m³/yr). Period: 1997–2001. (See p. 126.)

Chapter 8

The Water Footprint of Coffee and Tea Consumption

Growing environmental awareness has made people more frequently ask the question: What are the hidden natural resources in a product? Which natural resources, and how much of them, were needed in order to enable us to consume a certain product? In this chapter we will look at the water use behind coffee and tea. In the next chapter we will consider cotton, another big water consumer.

Coffee and tea consumption is possible through the use of natural and human resources in the producing countries. One of the natural resources required is water. There is a particular water need for growing the plant, but there is also a need for water to process the crop into the final product. When there is a transfer of a product from one place to another, there is little direct, physical transfer of water (apart from the water content of the product, which is quite insignificant in terms of volume). There is however a significant transfer of "virtual water" from the coffee- and tea-producing countries to the consuming countries. The consumers in the importing countries are indirectly employing the water in the producing countries. However, consumers generally have little idea of the resources needed to enable them to consume. In this chapter we assess the volume of water needed to allow the Dutch to drink coffee and tea. Though we focus on Dutch coffee and tea consumption, most findings will be applicable for consumers in other countries as well.

The roots of coffee consumption are probably in Ethiopia. The coffee tree is said to originate in the province of Kaffa (ICO, 2003). Coffee spread to the different parts of the world in the 17th and 18th centuries, the period of colonization. By the early 18th century the

Dutch colonies had become the main suppliers of coffee to Europe. Today people all over the world drink coffee. The importance of coffee to people cannot easily be overestimated. Coffee is of great economic importance to the producing, mostly developing, countries, and of considerable social importance to the consuming countries.

Tea is the dried leaf of the tea plant. The two main varieties of tea plant are *Camellia sinensis* and *Camellia assamica*. Indigenous to both China and India, the plant is now grown in many countries around the world. Tea was first consumed as a beverage in China sometime between 2700 BC and AD 220. The now traditional styles of green, black, and oolong teas first made an appearance in the Ming Dynasty in China (AD 1368–1644). Tea began to travel as a trade item as early as the fifth century, with some sources indicating Turkish traders bartering for tea on the Mongolian and Tibetan borders. Tea made its way to Japan late in the sixth century, along with another Chinese export product – Buddhism. By the end of the seventh century, Buddhist monks were planting tea in Japan. Tea first arrived in the west via overland trade into Russia. Certainly Arab traders had dealt in tea before this time, but no Europeans had a hand in tea as a trade item until the Dutch began an active and lucrative trade early in the 17th century. Dutch and Portuguese traders were the first to introduce Chinese tea to Europe. The Portuguese shipped it from the Chinese coastal port of Macao; the Dutch brought it to Europe via Indonesia. From Holland, tea spread relatively quickly throughout Europe. Although drunk in varying amounts and different forms, tea is the most consumed beverage in the world next to water. Tea is grown in over 45 countries around the world, typically between the tropics of Cancer and Capricorn (FAO, 2006e). We limit ourselves here to tea made from the real tea plant (i.e. *Camellia sinensis* and *C. assamica*). This excludes other sorts of "tea," made from other plants, such as "rooibos tea" (from a reddish plant grown in South Africa), "honey-bush tea" (related to rooibos tea and also grown in South Africa), "yerba mate" (from a shrub grown in some Latin American countries), and "herbal tea" (a catch-all term for drinks made from leaves or flowers from various plants infused in hot water).

As explained in Chapter 5, the impact of human consumption on the global water resources can be mapped with the concept of the "water footprint." The water footprint of a nation can be quantified as the total volume of freshwater that is used to produce the goods and services consumed by the inhabitants of the nation. The water footprint includes a spatial dimension by visualizing how consumers in one particular part of the world indirectly employ the water resources in various other parts of the world. In this chapter we will assess the global water footprint of Dutch society in relation to its coffee and tea consumption. As a first step, we estimate the virtual-water content of coffee and tea in each of the countries that export coffee or tea to the Netherlands. Next, we calculate the volumes of virtual-water flows entering and leaving the Netherlands in the period 1995–99 insofar as they are related to coffee and tea trade. Finally, we assess the volume of water needed to consume one cup of coffee or tea in the Netherlands. The water volume per cup multiplied by the number of cups consumed per year provides an estimate of the total Dutch annual water footprint related to coffee or tea consumption.

Virtual-Water Content of Coffee and Tea in Different Production Stages

The virtual-water content of coffee or tea is the volume of water required to produce one unit of coffee or tea, generally expressed in terms of cubic meters of water per ton of coffee or tea. This differs throughout the different stages of processing. In the case of coffee, the virtual-water content of fresh cherries is calculated based on the water use of the coffee plant (in m^3/ha) and the yield of fresh cherries (in ton/ha). After each processing step, the weight of the remaining product is smaller than the original weight. Following the methodology described in Chapter 2 (and in more detail in Appendix I) we define the "product fraction" in a certain processing step as the ratio of the weight of the resulting product to the weight of the original product. The virtual-water content of the resulting product (expressed in m^3/ton) is larger than the virtual-water content of the original product. It can be found by dividing the virtual-water content of the original product by the product fraction. If a particular processing step requires water (namely the processes of pulping, fermentation, and washing in the wet production method), the water needed (in m^3 per ton of original product) is added to the initial virtual-water content of the original product before translating it into a value for the virtual-water content of the resulting product. Figure 8.1 shows how the

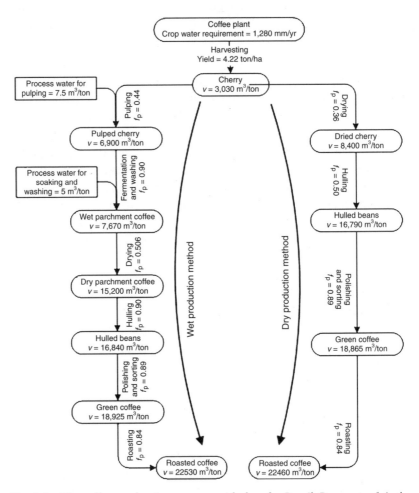

Fig. 8.1 The coffee-production process, with data for Brazil. Parameter f_p is the product fraction (ton of processed product per ton of root product) and v is the virtual-water content.

virtual-water content of coffee is calculated in its subsequent production stages, for both the wet and dry production methods. The numbers shown are those for Brazil. This scheme of calculation was adopted for all other countries with their respective crop water requirements and yields of fresh cherries. The product fractions and process water volumes are assumed to be constant across different coffee-producing countries.

In the case of tea, the virtual-water content of fresh leaves is calculated based on the water use of the tea plant (in m³/ha) and

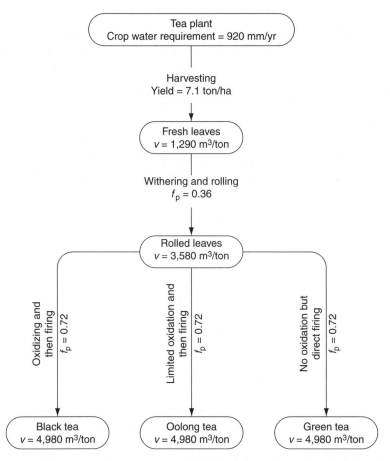

Fig. 8.2 The tea-production process, with data for India. Parameter f_p is the product fraction (ton of processed product per ton of root product) and v is the virtual-water content.

the yield of fresh leaves (in ton/ha). The virtual-water content at different stages of production is calculated using the same approach followed for coffee. Figure 8.2 shows how the virtual-water content of tea is calculated in its subsequent production stages for the case of the orthodox production of black tea in India.

From fresh cherries to green coffee the weight is reduced to about 16% of the original, due to removal of pulp and parchment, reduction in moisture content, and separating out of low-quality beans (GTZ, 2002a). The weight reduction occurs in steps. In the wet production method, only 44% of the fresh cherry remains after pulping (Bressani, 2003), 90% of the pulped cherry remains after fermentation and washing (Bressani, 2003), 51% of the wet parchment coffee remains after drying (GTZ, 2002c), and 80% of the dry parchment coffee remains after hulling, polishing, and sorting (GTZ, 2002a). In the dry production method, about 36% of the fresh cherry remains after drying (Hicks, 2001), 50% of the dried cherry remains after hulling (Hicks, 2001), and 89% of the hulled beans remain after polishing and sorting. From green coffee to roasted coffee there is another weight reduction, due to a decrease in moisture content. The remaining fraction after roasting is generally reported to be 84% of the original green coffee (Hicks, 2001; GTZ, 2002a; ICO, 2003).

The wet production method requires water for pulping, fermentation, and washing. The total amount of water needed ranges between 1 and 15 m^3 per ton of cherry (GTZ, 2002b). Here we crudely assume that 7.5 m^3 of water per ton of fresh cherry is needed in the pulping process and 5 m^3 of water per ton of pulped cherry is needed in the fermentation and washing process (Roast and Post, 2003). If we bring these two numbers into one denominator, this is equivalent to about 10 m^3 of water per ton of cherry. We will see later that the estimate for the total water needs for making coffee is not sensitive to these assumptions.

Table 8.1 shows the calculations for all coffee-producing countries from which the Netherlands imports coffee. These countries are together responsible for 84% of global coffee production. The differences between the dry and wet production methods in terms of total water needs are trivial. Most water is needed for growing the coffee plant. In the wet production method, only 0.34% of the total water need refers to process water.

In the case of tea, 10 kg of green shoots (containing 75–80% water) produce about 2.5 kg of dried tea. The overall remaining fraction after processing fresh tea-leaves into made tea is thus 0.25. The weight reduction occurs in two steps. Withering reduces moisture content by up to 70% and drying further reduces it to about 3% (Twinings, 2003). There is no weight reduction in the rolling and oxidation processes. In our study for this book, we have assumed that the remaining fraction after withering is 0.72 (ton of withered tea per ton of fresh leaves) and that the remaining fraction after firing is 0.36 (ton of black tea per ton of rolled leaves). The different

Table 8.1 Virtual-water content of coffee by country. Period: 1995–99.

Country	Production of green coffee* (ton/yr)	Yield of green coffee* (ton/ha)	Yield of fresh cherries (ton/ha)	Crop water requirement (mm/yr)	Virtual-water content (m³/ton)						
					Fresh cherries	Pulped cherries	Wet parchment coffee	Dry parchment coffee	Hulled beans	Green coffee	Roasted coffee
Brazil	13,70,000	0.68	4.22	1,277	3,030	6,880	7,670	15,160	16,840	18,930	22,530
Colombia	6,90,000	0.74	4.61	893	1,940	4,410	4,920	9,720	10,800	12,140	14,450
Indonesia	4,66,000	0.55	3.41	1,455	4,270	9,700	10,800	21,350	23,720	26,650	31,730
Vietnam	3,84,000	1.87	11.60	938	810	1,830	2,060	4,070	4,530	5,090	6,050
Mexico	3,29,000	0.46	2.88	1,122	3,900	8,860	9,870	19,500	21,670	24,350	28,990
Guatemala	2,40,000	0.90	5.60	1,338	2,390	5,430	6,060	11,970	13,300	14,940	17,790
Uganda	22,9000	0.84	5.25	1,440	2,740	6,230	6,950	13,730	15,250	17,140	20,400
Ethiopia	2,27,000	0.91	5.65	1,151	2,040	4,630	5,170	10,210	11,350	12,750	15,180
India	2,20,000	0.81	5.08	754	1,490	3,380	3,770	7,460	8,290	9,310	11,090
Costa Rica	1,57,000	1.47	9.14	1227	1340	3050	3,410	6,750	7,500	8,420	10,030
Honduras	1,55,000	0.78	4.87	1,483	3,040	6,920	7,710	15,240	16,940	19,030	22,650
El Salvador	1,38,000	0.85	5.28	1,417	2,690	6,100	6,810	13,450	14,940	16,790	19,990
Ecuador	1,21,000	0.32	1.98	1,033	5,230	11,880	13,220	26,130	29,030	32,620	38,830
Peru	1,16,000	0.61	3.80	994	2,610	5,940	6,620	13,080	14,540	16,340	19,450
Thailand	75,800	1.12	6.96	1,556	2,240	5,080	5,670	11,210	12,450	13,990	16,660
Venezuela	67,800	0.35	2.19	1,261	5,760	13,080	14,560	28,780	31,970	35,920	42,770
Nicaragua	65,400	0.73	4.55	1,661	3,650	8,290	9,240	18,260	20,290	22,800	27,140

(Continued)

Table 8.1 (Continued).

Country	Production of green coffee* (ton/yr)	Yield of green coffee* (ton/ha)	Yield of fresh cherries (ton/ha)	Crop water requirement (mm/yr)	Virtual-water content (m³/ton)						
					Fresh cherries	Pulped cherries	Wet parchment coffee	Dry parchment coffee	Hulled beans	Green coffee	Roasted coffee
Madagascar	63,200	0.33	2.04	1,164	5,700	12,940	14,400	28,450	31,610	35,520	42,290
Tanzania	44,500	0.38	2.38	1,422	5,960	13,560	15,090	29,810	33,130	37,220	44,310
Bolivia	22,600	0.94	5.84	1,093	1,870	4,260	4,760	9,400	10,440	11,730	13,970
Togo	14,400	0.34	2.12	1,409	6,640	15,100	16,800	33,200	36,890	41,450	49,340
Sri Lanka	11,100	0.68	4.22	1,426	3,380	7,680	8,560	16,910	18,790	21,120	25,140
Panama	10,700	0.41	2.55	1,294	5,070	11,520	12,820	25,340	28,160	31,630	37,660
Ghana	4,910	0.35	2.16	1,381	6,400	14,550	16,190	32,000	35,550	39,950	47,550
USA	2,920	1.24	7.74	938	1,210	2,750	3,090	6,100	6,770	7,610	9,060
Average†	—	0.80	4.53	1,195	2,820	6,410	7,150	14,100	15,700	17,600	21,000

* Source: FAO (2006e).

† Country data have been weighted on the basis of their share of green coffee to the global production, which is 6,202,000 ton/yr.

methods of processing fresh tea-leaves into black, green, or oolong tea turn out to be more or less equal when it comes to the remaining fraction (ton of made tea per ton of fresh tea-leaves). For that reason, we have not distinguished between different production methods when calculating the virtual-water content of tea in the different tea-producing countries. The calculation is made for black tea and assumed to be representative for green tea and oolong tea as well.

Table 8.2 presents the virtual-water content of tea in the different production steps for all tea-producing countries that export tea to the Netherlands. These countries are together responsible for 81% of global tea production. The global average virtual-water content of fresh tea-leaves is 2.7 m^3/kg. The average virtual-water content of made tea is 10.4 m^3/kg. The latter figure has been based on a calculation for black tea, but there would be hardly any difference for green or oolong tea, because the overall weight reduction in the case of these teas is similar to the weight reduction when producing black tea. Besides, it is worth noting here that black tea takes the largest share (78%) of the global production of tea.

Virtual-Water Flows Related to the Trade in Coffee and Tea

The volume of virtual water imported into the Netherlands (in m^3/yr) as a result of coffee or tea import can be found by multiplying the amount of product imported (in ton/yr) by the virtual-water content of the product (in m^3/ton). The virtual-water content of tea and coffee is taken from the exporting countries. The volume of virtual water exported from the Netherlands is calculated by multiplying the export quantity by the respective average virtual-water content of coffee and tea in the Netherlands. The latter is taken as the average virtual-water content of the coffee and tea imported into the Netherlands. The difference between the total virtual-water import and the total virtual-water export is the net virtual-water import to the Netherlands, an indicator for the total amount of water needed to allow the Dutch to drink coffee and tea.

Data on coffee and tea trade were taken from the PC-TAS database of the International Trade Centre in Geneva for the period 1995–99 (ITC, 2002). The total volumes of coffee and tea imported into the

Table 8.2 Virtual-water content of tea by country. Period: 1995–99.

Country	Production of made tea* (ton/yr)	Yield of made tea* (ton/ha)	Yield of fresh tea-leaves (ton/ha)	Crop water requirement (mm/yr)	Virtual-water content (m³/ton)		
					Fresh tea-leaves	Withered and rolled leaves	Made tea
India	7,94,000	1.84	7.10	917	1,290	3,580	4,980
China	6,49,000	0.73	2.80	1,205	4,300	12,000	16,600
Sri Lanka	2,69,000	1.41	5.45	1,731	3,170	8,820	12,200
Indonesia	1,60,000	1.43	5.51	1,769	3,210	8,920	12,400
Turkey	1,47,000	1.91	7.38	1,349	1,830	5,080	7,050
Japan	87,100	1.68	6.47	1,165	1,800	5,000	6,950
Argentina	53,100	1.40	5.39	1,286	2,390	6,630	9,210
Bangladesh	51,900	1.08	4.15	1,404	3,380	9,400	13,100
Tanzania	24,100	1.29	4.98	1,726	3,470	9,630	13,400
Uganda	20,400	1.12	4.32	1,746	4,050	11,200	15,600
South Africa	10,900	1.66	6.41	1,822	2,840	7,890	11,000
Brazil	6,750	1.84	7.11	1,550	2,180	6,060	8,410
Mauritius	2,210	2.15	8.31	1,548	1,860	5,180	7,190
Weighted mean	—	—	—	—	2,690	7,480	10,400

* Source: FAO (2006e).

Table 8.3   Coffee and tea trade in the Netherlands by product type. Period: 1995–99.

Product code in PC-TAS	Product	Import (ton/yr)	Export (ton/yr)
090111	Coffee, not roasted, not decaffeinated	1,35,400	7,250
090112	Coffee, not roasted, decaffeinated	5,330	731
090121	Coffee, roasted, not decaffeinated	22,000	7,230
090122	Coffee, roasted, decaffeinated	3,890	1,440
090210	Green tea in packages not exceeding 3 kg	225	51
090220	Green tea in packages exceeding 3 kg	936	17
090230	Black tea and partly fermented tea in packages not exceeding 3 kg	2,580	1,350
090240	Black tea and partly fermented tea in packages exceeding 3 kg	13,500	7,980

Source: ITC (2002). PC-TAS, Personal Computer Trade Analysis System.

Netherlands and the total volumes exported are presented in Table 8.3. The data are given for four different coffee products: non-decaffeinated non-roasted, decaffeinated non-roasted, non-decaffeinated roasted, and decaffeinated roasted coffee. The term "non-roasted coffee" in PC-TAS refers to what is generally called "green coffee." Some of the countries exporting coffee to the Netherlands do not grow coffee themselves. These countries import the coffee from elsewhere in order to further trade it. The trade database uses the different terminology "fermented" and "not fermented" tea for oxidized (black) and non-oxidized (green) tea, respectively.

The virtual-water import to the Netherlands as a result of coffee import is 3.0 billion m³/yr (Table 8.4). Brazil and Colombia together are responsible for 25% of this import. Other important sources are Guatemala (5%), El Salvador (5%), and Indonesia (4%), as shown in Map 12. A large proportion of the coffee import comes from the non-coffee-producing countries Belgium and Germany (34% in total). The virtual-water import to the Netherlands as a result of tea import in the period 1995–99 was 0.2 billion m³/yr on average (Table 8.5). Indonesia is the largest source, contributing 35% of the total import into the Netherlands. Other sources are China (21%), Sri Lanka (14%), Argentina (6%), India (5%), Turkey (3%), and Bangladesh (1%). There is also some import from within Europe: Germany (6%),

Table 8.4 Virtual-water balance of the Netherlands due to coffee trade. Period: 1995–99.

Origin	Virtual-water import (10^6 m³/yr)	Destination	Virtual-water export (10^6 m³/yr)
Belgium-Luxembourg	612	Belgium-Luxembourg	73.5
Brazil	426	UK	61.1
Germany	380	Germany	55.5
Colombia	324	France	38.7
Guatemala	159	Sweden	16.8
El Salvador	154	Spain	10.9
Indonesia	127	Denmark	9.9
Togo	99	USA	7.6
Tanzania	92	Russian Federation	6.9
Mexico	85	Italy	4.3
Costa Rica	75	Norway	4.1
Nicaragua	73	Finland	2.9
Peru	72	Netherlands Antilles	2.0
Honduras	48	Austria	1.6
India	36	Lithuania	1.5
France	34	Greece	1.4
Uganda	32	Czech Republic	1.4
Ecuador	19	Aruba	1.3
Italy	19	Portugal	1.2
Others	89	Turkey	1.0
		Others	10.3
Total	2,953	Total	314

Switzerland (4%), the UK (2%), and Belgium-Luxembourg (2%). It is difficult to trace back the original source of the coffee and tea imported from the countries that do not produce the crop themselves. In these cases we have taken the global average virtual-water content of the product from Tables 8.1 and 8.2.

The total import of green coffee over the period 1995–99 amounts to 141×10^3 ton/yr. The import of roasted coffee is 26×10^3 ton/yr. The average virtual-water content of coffee imported into the Netherlands is 17.1 m³ per kg of green coffee and 20.4 m³ per kg of roasted coffee. These figures are very close to the average global virtual-water content of green and roasted coffee, respectively. The total import of tea over the

Table 8.5 Virtual-water balance of the Netherlands due to tea trade. Period: 1995–99.

Origin	Virtual-water import (10^6 m^3/yr)	Destination	Virtual-water export (10^6 m^3/yr)
Indonesia	69.2	Germany	22.2
China	41.2	UK	18.6
Sri Lanka	28.2	Russian Federation	16.5
Argentina	12.0	Switzerland	8.1
Germany	11.5	USA	6.6
India	8.6	Italy	4.5
Switzerland	7.2	France	4.3
Turkey	6.1	Belgium-Luxembourg	2.7
UK	3.5	Saudi Arabia	2.4
Belgium-Luxembourg	3.2	Denmark	2.2
Bangladesh	1.9	Canada	1.6
Tanzania	1.0	Austria	1.5
Brazil	0.9	Finland	1.4
Others	2.1	Others	14.6
Total	197	Total	107

period 1995–99 amounts to 17×10^3 ton/yr. The average virtual-water content of tea imported into the Netherlands is 11.4 m^3 per kg of made tea. This figure is very close to the global average virtual-water content of made tea, which is 10.4 m^3 per kg.

The total virtual-water export from the Netherlands as a result of coffee export is 0.3 billion m^3/yr. The largest importers of virtual water from the Netherlands through coffee are: Belgium-Luxembourg (23%), the UK (20%), Germany (18%), and France (12%). The total virtual-water export from the Netherlands as a result of tea export is 0.1 billion m^3/yr. The largest importers of virtual water from the Netherlands through tea are: Germany (21%), the UK (17%), the Russian Federation (15%), Switzerland (8%), the USA (6%), Italy (4%), France (4%), and Belgium-Luxembourg (3%). Virtual-water exports from the Netherlands as a result of coffee and tea export are presented in Tables 8.4 and 8.5, respectively.

Coffee and tea import are responsible for about 5% of the total gross virtual-water import into the Netherlands related to the import of agricultural products (see Appendix II). The Dutch water footprint related to tea consumption is much smaller than the Dutch water footprint related to coffee consumption, due to the facts that the Dutch consume relatively large amounts of coffee and that tea has a much lower virtual-water content than coffee.

The Water Needed to Drink a Cup of Coffee or Tea

The quantity of roasted coffee per cup of coffee varies from 5 to 10 g. For the calculation of the virtual-water content of a standard cup of coffee, we assume here a figure of 7 g of roasted coffee. Based on the average virtual-water content of roasted coffee ($20.4 \, m^3/kg$), one cup of coffee requires about 140 liters of water in total. A standard cup of coffee is 125 ml, which means that we need more than 1,100 drops of water for producing one drop of coffee. For making 1 kg of soluble coffee powder, one needs 2.3 kg of green coffee (Rosenblatt et al., 2003). That means that the virtual-water content of instant coffee is about $39,400 \, m^3/ton$. This is much higher than in the case of roasted coffee, but for making one cup of instant coffee one needs a relatively small amount of coffee powder (about 2 g). Surprisingly, the virtual-water content of a cup of instant coffee is thus lower than the virtual-water content of a cup of standard coffee.

A cup of tea typically requires 1.5–3 g of processed tea. We assume here 3 g of processed tea (either black, green, or oolong tea) for a cup of normal-strength tea and 1.5 g for "weak" tea. This is equivalent to 34 liters of water per cup of standard tea and 17 liters per cup of weak tea. Thus one consumes about four times more water if one chooses a cup of coffee instead of a cup of tea. With a standard cup of tea of 250 ml, we need about 136 drops of water for producing one drop of tea. The results as presented in Table 8.6 for the Netherlands are fairly representative for the global average, so the figures can be cited in more general terms as well.

The Water Footprint of Coffee and Tea Consumption

With an average of three cups of coffee a day per person, the Dutch community has a coffee-related water footprint of 2.6 billion m^3 of water per year. This is the volume of water appropriated in the

Table 8.6 Virtual-water content of a cup of coffee or tea.

		Virtual-water content of the dry ingredient (m^3/kg)	One cup of coffee or tea		
			Dry product content (g/cup)	Real-water content (liter/cup)	Virtual-water content (liter/cup)
Coffee	Standard cup of coffee	20.4	7	0.125	140
	Weak coffee	20.4	5	0.125	100
	Strong coffee	20.4	10	0.125	200
	Instant coffee	39.4	2	0.125	80
Tea	Standard cup of tea	11.4	3	0.250	34
	Weak tea	11.4	1.5	0.250	17

coffee-producing countries. The total volume is equivalent to 36% of the annual flow of the Meuse, a Dutch river. The Dutch people account for 2.4% of the world's coffee consumption. In total, the world population requires about 110 billion m^3 of water per year in order to be able to drink coffee. This is equivalent to 15 times the annual runoff of the Meuse, or 1.5 times the annual runoff of the Rhine.

The water footprint of Dutch tea consumption is 90 million m^3 per year. Dutch people account for only 0.28% (7.8×10^3 ton/yr) of the world's tea consumption (2.82 million ton/yr). The world population requires about 30 billion m^3 of water per year in order to be able to drink tea.

The water needed to drink coffee or tea in the Netherlands is not actually Dutch water, because the crops are not produced in the Netherlands. Coffee is produced in Latin America (Brazil, Colombia, Guatemala, El Salvador, Mexico, Costa Rica, Nicaragua, Peru, and Honduras), Africa (Togo, Tanzania, and Uganda), and Asia (Indonesia and India). The most important sources are Brazil and Colombia. Tea is produced in Southeast Asia (Indonesia, China, Sri Lanka, India, and Bangladesh) and some other countries in different parts of the world

(Argentina, Turkey, Brazil, Tanzania, and South Africa). As mentioned, there are also coffee and tea imports from countries that do not produce coffee or tea themselves, such as Germany and Belgium. These are merely intermediate countries, where coffee or tea is simply transited or upgraded (e.g. through blending or being made into brand names to gain higher economic returns).

The consumption of coffee and tea in the Netherlands has positive impacts on the economies of the producing countries. It generates economic benefits to the producing countries (which are mostly developing countries) with the use of a resource (rainwater) that has relatively low opportunity cost. Although rainwater appropriated for coffee or tea production will often have no alternative use (e.g. production of another crop or natural forest) that might provide higher economic or social return, the economic value of rainwater should be included in the price of the product (Hoekstra et al., 2001, 2003; Albersen et al., 2003). In the exceptional cases where irrigation is applied, it is even more important to pass on the economic cost of the water to the consumers of the coffee or tea, because competition over irrigation water is generally more severe than competition over rainwater. In practice, however, the economic cost of rainwater is never included in the price of the product, while the economic cost of irrigation water is usually only partly incorporated. The reason is that irrigation water is generally heavily subsidized (Cosgrove and Rijsberman, 2000).

The volume of water needed to make coffee and tea depends particularly on the climate at the place of production and the yields per hectare that are obtained. The latter depends partly on the climatic conditions, but also on soil conditions and management practices. For the overall water needs, it makes hardly any difference whether coffee is produced with the dry or the wet production process, because the water used in the wet production process is only a very small fraction (0.34%) of the water used to grow the coffee plant. However, this relatively small amount of water can be and actually often is a problem, because this is water that has to be obtained from surface or groundwater, which is generally scarcer than rainwater (i.e. competition is greater). Besides, the wastewater from the coffee factories is often heavily polluted (GTZ, 2002b). In current practice, coffee as bought by final consumers neither includes in its price the economic costs of water inputs and water impacts, nor reveals qualitative information about those costs on its label. We recommend further exploration of the economic, social, and environmental costs of water use, and the development of practical models for passing on the costs of water inputs and impacts to the consumer through the price of coffee in the shop.

Chapter 9

The Water Footprint
of Cotton Consumption

In the previous chapters we have introduced the water-footprint concept, shown the actual water footprints for all nations of the world, looked in more detail at the cases of the Netherlands, Morocco, and China, and considered in detail the global water footprint of Dutch coffee and tea consumption. In the current chapter we focus on another water-consuming commodity: cotton. We thereby take a global perspective: we assess the impact of worldwide consumption of cotton products on the water resources in the cotton-producing countries.

In order to visualize the impact of human consumption on the global water resources we use again the concept of the water footprint. In the previous water-footprint chapters we restricted ourselves to the quantification of production-related evaporation of groundwater, surface water, and rainwater. The effect of pollution was included to a limited extent only, namely by including the polluted return flows in the industrial and domestic sectors. The effect of pollution on water resources was underestimated because one cubic meter of wastewater generally pollutes many more cubic meters of water in the natural system. The precise multiplication factor depends on the concentration of pollutants in the waste flow. In this chapter we apply an improved method of quantifying the impacts of pollution. The approach is to quantify the water volumes required to dilute waste flows to such an extent that the quality of the water meets agreed water quality standards. The rationale for including this water component in the definition of the water footprint is similar to the rationale for including the land area needed for uptake of anthropogenic carbon dioxide emissions in the definition of the ecological footprint. Land and water function not only as resource bases, but also as systems for waste assimilation. We realize that the method to translate the impacts of pollution into water requirements as applied here can potentially invoke a similar debate as is being held about the methods applied to translate the impacts of carbon dioxide emissions into land requirements (see for example Van den Bergh and Verbruggen, 1999; Van Kooten and Bulte, 2000). We would welcome such a debate, because of the societal need for proper natural resources accounting systems on the one hand and the difficulties in achieving the required scientific rigor in the accounting procedures on the other.

Some of the earlier studies on the impacts of cotton production were limited to the impacts in the industrial stage only (e.g. Ren, 2000), leaving out the impacts in the agricultural stage. Other cotton impact studies use the method of life cycle analysis and thus include all stages of production, but these studies are focused on methodology rather than the quantification of the impacts (e.g. Proto et al., 2000; Seuring, 2004). Earlier studies that have gone in the direction we are aiming at here are the background studies for the cotton initiative of the World Wide Fund for Nature (Soth et al., 1999; De Man, 2001). In this chapter, however, we will synthesize the various impacts of cotton on water in one comprehensive indicator, the water footprint, and we will introduce the spatial dimension by showing how water footprints of some nations particularly impact on other parts of the world.

Cotton is the most important natural fiber used in the textile industry worldwide. Today, cotton makes up about 40% of textile production, while synthetic fibers make up about 55% (Soth et al., 1999; Proto et al., 2000). During the period 1997–2001, international trade in cotton products constituted 2% of the global merchandise trade value.

The impacts of cotton production on the environment are easily visible and have different aspects. On the one hand there are the effects of water depletion, on the other hand the effects on water quality. In many of the major textile processing areas, those living close to the river downstream can see from the river what was the latest color applied in the upstream textile industry. The Aral Sea in Central Asia is the most famous example of the effects of water abstractions for irrigation. In the period 1960–2000, the Aral Sea lost approximately 60% of its area and 80% of its volume (Glantz,

1998; Hall et al., 2001; Pereira et al., 2002; UNEP, 2002; Loh and Wackernagel, 2004) as a result of the annual abstractions of water from the Amu Darya and the Syr Darya – the rivers which feed the Aral Sea – to grow cotton in the desert.

About 53% of the global cotton field is irrigated, resulting in 73% of the global cotton production (Soth et al., 1999). Irrigated cotton is grown mainly in the Mediterranean and other warm climatic regions, where freshwater is already in short supply, and in other dry regions such as Egypt, Uzbekistan, and Pakistan. Cotton in the Xinjiang province of China is entirely irrigated whereas in Pakistan and the north of India a major portion of the crop water requirements of cotton is met by supplementary irrigation. As a result, in Pakistan already 31% of all irrigation water is drawn from ground water and in China the extensive use of freshwater has caused falling water tables (Soth et al., 1999). Nearly 70% of the world's cotton crop production is from China, the USA, India, Pakistan, and Uzbekistan (USDA, 2004). Most of the cotton productions rely on a furrow irrigation system. Sprinkler and drip systems are also adopted as an irrigation method in water-scarce regions. However, only about 0.7% of land in the world is irrigated by this method (Postel, 1992).

Green, Blue, and Gray Water

From field to end product, cotton passes through a number of distinct production stages with different impacts on water resources. These stages of production are often carried out at different locations, and consumption can take place at yet another location. For instance, Malaysia does not grow cotton, but imports raw cotton from China, India, and Pakistan for processing in the textile industry and exports cotton clothes to the European market. For this reason the impacts of consumption of a final cotton product can only be found by tracing the origins of the product. The relation between the production stages and their impacts on the environment is shown in Fig. 9.1.

Although the chain from cotton growth to final product can involve several distinct steps, there are two major stages: the agricultural stage (cotton production at field level) and the industrial stage (processing of seed cotton into final cotton products). In the first stage, there are three types of impact: evaporation of infiltrated rainwater for

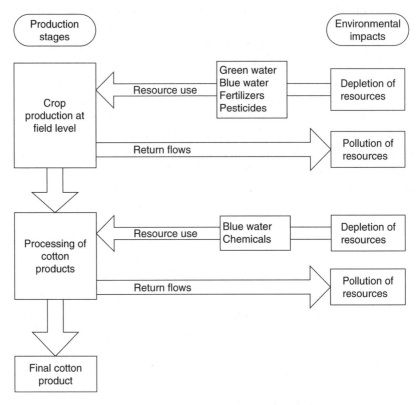

Fig. 9.1 The impact of cotton production on natural resources.

cotton growth, withdrawal of ground- or surface water for irrigation, and water pollution due to the leaching of fertilizers and pesticides. Based on Falkenmark (2003), we use the terms "green water use" and "blue water use" to distinguish between the two different types of water source (either rainwater or ground-/surface water). "Green water use" is quantitatively defined as the volume of rainwater that evaporated from the cotton field. "Blue water use" refers to the volume of irrigation water that evaporated from the field. The latter definition provides a conservative estimate of blue water use, because the volume of withdrawal from ground- or surface water for irrigation is larger than the volume that ultimately evaporates from the field. The difference consists of "losses" due to infiltration or evaporation during transport and application. These "losses" however are available again insofar as they concern infiltration losses. The impact on water quality is quantified here and made comparable to the impacts of water use by translating the volumes of emitted chemicals into the dilution volume necessary to assimilate the pollution. The term "gray water footprint" refers to this required dilution volume. In the industrial stage, there are two major impacts on water: evaporation of process water abstracted from surface or groundwater (blue water use), and pollution of water as a result of the waste flows from the cotton processing industries. The latter is again translated into a dilution volume requirement.

The Virtual-Water Content of Seed Cotton

The virtual-water content of seed cotton (m^3/ton) was calculated as the ratio of the volume of water (m^3/ha) used during the entire period of crop growth to the corresponding crop yield (ton/ha). The volume of water used to grow crops in the field has two components: effective rainfall (green water) and irrigation water (blue water). The CROP-WAT model of the Food and Agriculture Organization was used to estimate the effective rainfall and the irrigation requirements for each country (Allen et al., 1998; FAO, 2006b). The climate data were taken from FAO (2006c,d) for the most appropriate climate stations (USDA/NOAA, 2005a) located in the major cotton-producing regions of each country. The actual irrigation water use is taken as equal to the irrigation requirements as estimated with the CROPWAT model for those countries where the whole harvesting area is reportedly irrigated. In the countries where only a certain fraction of the harvesting area is irrigated, the actual irrigation water use is taken as equal to this fraction times the irrigation requirements.

The "green" virtual-water content of the crop was estimated as the green component in crop water use (the minimum of crop water requirement and effective rainfall) divided by the crop yield (see Appendix I). Similarly, the "blue" virtual-water content of the crop was taken as equal to the blue component in crop water use (the minimum of irrigation water requirement and effective irrigation) divided by the crop yield. The total virtual-water content of seed cotton is the sum of the green and blue components, calculated separately for the 15 largest cotton-producing countries. These countries contribute nearly 90% of the global cotton production (Table 9.1). For the remaining countries the global average virtual-water content of seed cotton was assumed.

Table 9.1   The top 15 seed cotton-producing countries. Period: 1997–2001.

Country	Average production of seed cotton* (10^3 ton/yr)	Planting period[†]	Yield of seed cotton* (ton/ha)
China	13,604	April/May	3.16
USA	9,700	March/May	1.86
India	5,544	April/May/July	0.62
Pakistan	5,160	May/June	1.73
Uzbekistan	3,342	April	2.24
Turkey	2,200	April/May	3.12
Australia	1,777	October/November	3.74
Brazil	1,613	October	2.06
Greece	1,253	April	3.02
Syria	1,017	April/May	3.92
Turkmenistan	954	March/April	1.72
Argentina	712	October/December	1.16
Egypt	710	February/April	2.39
Mali	463	May/July	1.03
Mexico	454	April	2.98
Others	5,939	—	—
Global total	54,444	—	—

* *Source:* FAO (2006e); data on cottonseed have been calculated back to seed cotton.
[†] *Sources:* Cotton Australia (2005), UNCTAD (2005a), FAO (2006a).

In the 15 largest cotton-producing countries, the major cotton-producing regions were identified (Table 9.2) so that the appropriate climate data could be selected. For regions with more than one climate station, the data for the relevant stations were equally weighted assuming that the stations represent equal-sized cotton-producing areas. National average crop water requirements were calculated on the basis of the respective contribution of each region to the national production.

The calculated national average crop water requirements for the 15 largest cotton-producing countries are presented in Table 9.3. The table also shows effective rainfall in the cotton areas, irrigation requirements, actual irrigation, and finally the blue and green virtual-water content of seed cotton. The global average virtual-water content of seed cotton, excluding gray water, is 3,644 m^3/ton. With a total seed-cotton production in the world of 54 million ton per year, this means that the global volume of blue and green water use for cotton

Table 9.2 Major regions of cotton production within the main cotton-producing countries.

Country	Major cotton harvesting regions and their share of the national harvesting area
Argentina	Chaco (85%)
Australia	New South Wales (77%) and Queensland (23%)
Brazil	Parana (43%), São Paulo (21%), Bahia (8%), Mato Grosso (5%), Minas Gerais (5%), Goias (4%), and Mato Gross do Sul (4%)
China	Xinjiang (21.5%), Henan (16.6%), Jiangsu (11.5%), Hubei (11.4%), Shandong (10%), Hebei (6.7%), Anhui (6.4%), Hunan (5.2%), Jiangxi (3.3%), Sichuan (2.3%), Shanxi (1.7%), and Zhejiang (1.3%)
Egypt	Cairo (85%)
Greece	Thessaly (51%), East Macedonia (27%), and Central Macedonia (14%),
India	Punjab (18%), Andhra Pradesh (14%), Gujarat (14%), Maharastra (13%), Haryana (10%), Madhya Pradesh (10%), Karnataka (8%), Rajasthan (8%), and Tamil Nadu (4%)
Mali	Segou (85%)
Mexico	Baja California, Chihuahua, and Coahuila
Pakistan	Punjab (85%) and Sindh (15%)
Syria	Al Hasakah (33%), Ar Raqqah (33%), and Dayr az Zawr (33%)
Turkey	Southeastern Anatolia (45%), Aegean/Izmir (33.6%), Cukurova (20.2%), and Antalya (1.2%)
Turkmenistan	Ahal (85%)
USA	Texas (33.7%), Mississippi–Missouri–Tennessee–Arkansas–Louisiana (27.7%), California–Arizona (14.3%), Georgia (9.6%), and North Carolina (5.4%)
Uzbekistan	Fergana (85%)

Source: USDA/NOAA (2005b).

crop production is 200 billion m^3/yr, with a nearly equal share of green and blue water.

The water use for cotton production differs considerably over the different countries. Climatic conditions for cotton production are least attractive in Syria, Egypt, Turkmenistan, Uzbekistan, and Turkey,

Table 9.3 The blue and green virtual-water content of seed cotton. Period: 1997–2001.

Country	Crop water requirement (mm)	Effective rainfall (mm)	Irrigation requirement (mm)	Irrigated share of area* (%)	Total water use (mm)		Virtual-water content of seed cotton† (m³/ton)	
					Blue	Green	Blue	Green
Argentina	877	615	263	100	263	615	2,307	5,394
Australia	901	322	579	90	521	322	1,408	870
Brazil	606	542	65	15	10	542	46	2,575
China	718	397	320	75	240	397	760	1,258
Egypt	1,009	0	1,009	100	1,009	0	4,231	0
Greece	707	160	547	100	547	160	1,808	530
India	810	405	405	33	134	405	2,150	6,512
Mali	993	387	606	25	151	387	1,468	3,750
Mexico	771	253	518	95	492	253	1,655	852
Pakistan	850	182	668	100	668	182	3,860	1,054
Syria	1,309	34	1,275	100	1,275	34	3,252	88
Turkey	963	90	874	100	874	90	2,812	288
Turkmenistan	1,025	69	956	100	956	69	5,602	407
USA	516	311	205	52	107	311	576	1,673
Uzbekistan	999	19	981	100	981	19	4,377	83
Average	—	—	—	—	—	—	1,818	1,827

* Sources: Gillham et al. (1995), FAO (1999), CCI (2005), Cotton Australia (2005).
† Calculated by dividing the total water use by the seed-cotton yield.

because evaporative demand in these countries is very high (1,000–1,300 mm) while effective rainfall is very low (0–100 mm). The shortage of rain in these countries has been solved by irrigating the full harvesting area. Resulting yields vary from world-average (Turkmenistan) to very high (Syria and Turkey). Climatic conditions for cotton production are most attractive in the USA and Brazil. Evaporative demand is low (500–600 mm), so that vast areas can be grown without irrigation. Yields are a little above world-average. India and Mali are noteworthy in producing cotton under high evaporative water demand (800–1,000 mm), short-falling effective rainfall (400 mm), and using partial irrigation only (between a quarter and a third of the harvesting area), resulting in relatively low overall yields.

The average virtual-water content of seed cotton in the various countries gives a first rough indication of the relative impacts of the various production systems on water. Cotton from India, Argentina, Turkmenistan, Mali, Pakistan, Uzbekistan, and Egypt is most water intensive. Cotton from China and the USA on the other hand is least water intensive. Since blue water generally has a much higher opportunity cost than green water, it makes sense to look particularly at the blue virtual-water content of cotton in the various countries. China and the USA still show a positive picture in this comparative analysis. Brazil also appears in a positive light now, due to the acceptable yields under largely rain-fed conditions. The blue virtual-water content and thus the impact per unit of cotton production is highest in Turkmenistan, Uzbekistan, Egypt, and Pakistan, followed by Syria, Turkey, Argentina, and India.

It is interesting to compare neighboring countries such as Brazil and Argentina and India and Pakistan. Cotton from Brazil is preferable to cotton from Argentina from a water resources point of view because growth conditions are better in Brazil (smaller irrigation requirements) and, even despite the fact that the cotton harvesting area in Argentina is fully irrigated (compared with 15% in Brazil), the yields in Argentina are only half those in Brazil. Similarly, cotton from India is to be preferred over cotton from Pakistan – again from a water resources point of view only – because the effective rainfall in Pakistan's cotton harvesting area is low compared with that in India and the harvesting area in Pakistan is fully irrigated. Although India achieves very low cotton yields per hectare, the blue water requirements per ton of product are much lower in India than in Pakistan.

The Virtual-Water Content of Cotton Products

The different processing steps that transform the cotton plant through various intermediate products to some final products are shown in Fig. 9.2. The virtual-water content of seed cotton is attributed to its products following the method described in Appendix I. Thus the virtual-water content of each processed cotton product was calculated based on the product fraction (ton of crop product obtained per ton of primary crop) and the value fraction (market value of the crop product divided by the aggregated market value of all crop products derived from one primary crop). The product fractions were taken from the commodity trees in FAO (2003b) and UNCTAD (2005b). The value fractions were calculated based on the market prices of the various products. The global average market prices of the cotton products were calculated from ITC (2004). In calculating the virtual-water content of fabric, the process water volumes for bleaching, dying, and printing were added ($30\,\mathrm{m}^3$ per ton for bleaching, $140\,\mathrm{m}^3$ per ton for dying, and $190\,\mathrm{m}^3$ per ton for printing). Additional water

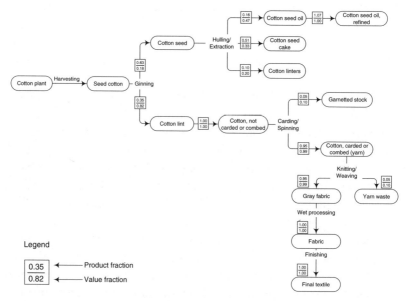

Fig. 9.2 The product tree for cotton.

is used in the finishing step (140 m^3 per ton). The process water volumes have to be understood as rough average estimates, because the actual water use varies considerably among different techniques (Ren, 2000).

Impact on Water Quality in the Crop Production Stage

Cotton production affects water quality in both the growing and the processing stages. The impact in the first stage depends on the amounts of nutrients (nitrogen, phosphorus, potash, and other minor nutrients) and pesticides that leach out of the plant root zone, thus contaminating ground- and surface water. In some cases, accumulation of chemicals in the soil (phosphorus) or the food chain (pesticides) is of concern as well. Most of the pesticides applied get into either groundwater or surface water bodies. Only 2.4% of the world's arable land is planted with cotton, yet cotton accounts for 24% of the world's insecticide market and 11% of the sale of global pesticides (WWF, 2003). N-fertilizer added to the field is partly taken up by the plant, partly transformed through denitrification into N$_2$ that dissipates into the atmosphere, and partly leaches out into groundwater or is washed away through surface runoff. In water bodies, high nitrogen concentrations can lead to problems of algal growth and increased cost of purification for drinking water.

About 60% of the total nitrogen applied is removed from the field in the form of harvested seed cotton (CRC, 2004). Silvertooth et al. (2001) estimate that out of the total nitrogen applied about 20% leaves the field through leaching to the groundwater, surface runoff, or denitrification to the atmosphere. Here we assume that the quantity of N that reaches free flowing water bodies is 10% of the applied rate, presuming a steady state balance at the root zone in the long term. The effect of other nutrients, pesticides, and herbicides used in cotton farming on the environment has not been analyzed.

The total volume of water required per ton N is calculated considering the volume of nitrogen leached (ton/ton) and the maximum allowable concentration in the free flowing surface water bodies. The standard recommended by EPA (2005) for nitrate in drinking water is 10 milligrams per liter (measured as nitrogen) and has been taken to calculate the necessary dilution water volume. This is a conservative approach, since natural background concentration of N in the water used for dilution has been assumed negligible.

We have used the average rate of fertilizer application for the year 1998 as reported by IFA et al. (2002). The total volume of fertilizer applied is calculated based on the average area of cotton harvesting for the period concerned. In this way we estimate that the 15 countries applied a total of 3 million tons of nitrogen per year on their cotton fields, of which 0.3 million tons leached to the groundwater or entered surface water through runoff (Table 9.4). Diluting this amount of nitrogen to acceptable concentrations required 30 billion m^3 of water per year. Given the cotton production in the 15 countries, this means an average gray water footprint of 620 m^3 per ton of seed cotton. This is equivalent to 1,500 m^3 of water per ton of final textile.

Impact on Water Quality in the Processing Stage

The average volumes of water used in wet processing (bleaching, dying, and printing) and the finishing stage are 360 m^3/ton and 136 m^3/ton of cotton textile, respectively (USEPA, 1996). The biological oxygen demand (BOD), chemical oxygen demand (COD), total suspended solids (TSS), and total dissolved solids (TDS) in the effluent from a typical textile industry are given by UNEP IE (1996) and presented in Table 9.5. The maximum permissible limits for effluents to be discharged into surface and groundwater bodies were taken from the guidelines set by the World Bank (1999).

As the maximum limits for different pollutants are different, the volume of water required to meet the desired level of dilution will be different per pollutant category in each production stage. Per production stage, the pollutant category that requires most dilution water has been taken as indicative for the total dilution water requirement (Table 9.6). The gray water footprint of final textile due to the use of chemicals during processing appears to be 880 m^3/ton, assuming that wet processing and finishing are carried out at different places.

The blue, green, and gray virtual-water content of different cotton products for the major cotton-producing countries is presented in Table 9.7. The gray virtual-water content refers to the volume of water necessary to dilute the fertilizer-enriched return flows from the cotton plantations and the polluted return flows from the processing industries. The overall water use for some specific consumer products is

Table 9.4 Fertilizer application to cotton fields and the volume of water that would be required to dilute the fertilizers leached to water bodies. Period: 1997–2001.

Country	Average fertilizer application rate* (kg/ha)			Total fertilizer applied (ton/yr)			Nitrogen leached to water bodies† (ton/yr)	Dilution water required‡ (10^6 m³/yr)	Gray virtual-water content of seed cotton (m³/ton)
	N	P_2O_5	K_2O	N	P_2O_5	K_2O			
Argentina	40	5	—	25,009	3,126	—	2,501	250	351
Australia	121	20	12.4	58,087	9,601	5,953	5,809	581	327
Brazil	40	50	50	30,674	38,342	38,342	3,067	307	190
China	120	70	25	516,637	301,372	107,633	51,664	5,166	380
Egypt	54	57	57	16,076	16,969	16,969	1,608	161	226
Greece	127	39	3.5	52,630	16,162	1,450	5,263	526	420
India	66	28	6	588,675	249,741	53,516	58,868	5,887	1,062
Mali	35	—	—	15,710	—	—	1,571	157	339
Mexico	120	30	—	18,315	4,579	—	1,831	183	404
Pakistan	180	28	0.4	536,720	83,490	1,193	53,672	5,367	1,040
Syria	50	50	—	12,964	12,964	—	1,296	130	128
Turkey	127	39	3.5	89,927	27,615	2,478	8,993	899	409
Turkmenistan	210	45	1.2	117,495	25,178	671	11,750	1,175	1,231

(*Continued*)

Table 9.4 (*Continued*).

Country	Average fertilizer application rate* (kg/ha)			Total fertilizer applied (ton/yr)			Nitrogen leached to water bodies[†] (ton/yr)	Dilution water required[‡] (10^6 m^3/yr)	Gray virtual-water content of seed cotton (m^3/ton)
	N	P$_2$O$_5$	K$_2$O	N	P$_2$O$_5$	K$_2$O			
USA	120	60	85	625,544	312,772	443,094	62,554	6,255	645
Uzbekistan	210	45	1.2	313,274	67,130	1,790	31,327	3,133	937
Average[§]	91	35	20	—	—	—	—	—	622
Total	—	—	—	3,017,737	1,169,041	673,090	301,774	30,177	—

* *Source*: IFA et al. (2002). For Uzbekistan, Mali, and Turkey, the fertilizer application rate has been taken from Turkmenistan, Nigeria, and Greece, respectively.

[†] Assuming that 10% of the nitrogen applied leaches to the natural water system (see text).

[‡] Calculated as the volume of nitrogen leached divided by the nitrogen concentration standard of 10 mg/l.

[§] The global average fertilizer application rate has been calculated from the country-specific rates, weighted on the basis of the share of a country in the global area of cotton production.

Table 9.5 Wastewater characteristics at different stages of processing cotton textiles and permissible limits for discharge into water bodies.

Process	Wastewater volume* (m³/ton)	Pollutants[†] (kg/ton of textile product)			
		BOD	COD	TSS	TDS
Wet processing	360	32	123	25	243
Bleaching	30	5	13	—	28
Dying	142	6	24	—	180
Printing	188	21	86	25	35
Finishing	136	6	25	12	17
Total	496	38	148	37	260
Permissible limits (mg/l)[‡]	—	50	250	50	—

* *Source:* USEPA (1996).
† *Source:* UNEP IE (1996). BOD, biological oxygen demand; COD, chemical oxygen demand; TSS, total suspended solids; TDS, total dissolved solids.
‡ *Source:* World Bank (1999).

Table 9.6 Volume of water necessary to dilute pollution at each production stage.

Stage of production	Volume of water per pollutant category (m³/ton of cotton textile)			Dilution water volume (m³/ton)
	BOD	COD	TSS	
Wet processing	640	492	500	640
Finishing	120	100	240	240
Wet processing and finishing carried out in the same place	760	592	740	760
Wet processing and finishing carried out in different places	—	—	—	880

shown in Table 9.8. One pair of jeans for example consumes about 12,000 liters of water on average: 4,900 liters of water from ground- and surface water sources, about 4,500 liters of rainwater, and 2,400 liters of water for dilution of pollutants. The last figure is the sum of the

Table 9.7 Virtual-water content of cotton products at different stages of production for the major cotton-producing countries (m^3/ton).

Country	Cotton lint			Gray fabric			Fabric			Final textile		
	Blue	Green	Gray	Blue	Green	Gray	Blue	Green	Gray	Blue	Green	Gray
Argentina	5,385	12,589	819	5,611	13,118	854	5,971	13,118	1,494	6,107	13,118	1,734
Australia	3,287	2,031	763	3,425	2,116	795	3,785	2,116	1,435	3,921	2,116	1,675
Brazil	107	6,010	444	112	6,263	462	472	6,263	1,102	608	6,263	1,342
China	1,775	2,935	886	1,849	3,059	924	2,209	3,059	1,564	2,345	3,059	1,804
Egypt	9,876	0	528	10,291	0	551	10,651	0	1,191	10,787	0	1,431
Greece	4,221	1,237	980	4,398	1,289	1,021	4,758	1,289	1,661	4,894	1,289	1,901
India	5,019	15,198	2,478	5,230	15,837	2,582	5,590	15,837	3,222	5,726	15,837	3,462
Mali	3,427	8,752	792	3,571	9,120	825	3,931	9,120	1,465	4,067	9,120	1,705
Mexico	3,863	1,990	942	4,026	2,073	982	4,386	2,073	1,622	4,522	2,073	1,862
Pakistan	9,009	2,460	2,428	9,388	2,563	2,530	9,748	2,563	3,170	9,884	2,563	3,410
Syria	7,590	204	298	7,909	213	310	8,269	213	950	8,405	213	1,190
Turkey	6,564	672	954	6,840	701	994	7,200	701	1,634	7,336	701	1,874
Turkmenistan	13,077	951	2,873	13,626	991	2,994	13,986	991	3,634	14,122	991	3,874
USA	1,345	3,906	1,505	1,401	4,070	1,569	1,761	4,070	2,209	1,897	4,070	2,449
Uzbekistan	10,215	195	2,188	10,644	203	2,280	11,004	203	2,920	11,140	203	3,160
Global average	4,240	4,260	1,450	4,420	4,440	1,510	4,780	4,440	2,150	4,900	4,450	2,400

Table 9.8   Global average virtual-water content of selected consumer products.

Consumer product	Standard weight (g)	Virtual-water content (liters)			
		Blue	Green	Gray	Total
1 pair of jeans	1,000	4,900	4,450	2,400	11,800
1 single bed sheet	900	4,400	4,000	2,150	10,600
1 t-shirt	250	1,230	1,110	600	2,900
1 diaper	75	370	330	180	880
1 cotton bud	0.33	1.6	1.5	0.8	4

water volume required to dilute fertilizers leached from the cotton field to the natural water system (1,500 liters) and the water volume needed to dilute chemicals used in the processing industry (880 liters).

International Virtual-Water Flows

In order to assess the water footprint of cotton consumption in a country we need to know the use of domestic water resources for domestic cotton growth or processing, as well as the water use associated with the import and export of raw cotton or cotton products. Virtual-water flows between nations (m³/yr) are calculated by multiplying cotton trade flows (ton/yr) by their associated total virtual-water content (m³/ton). In calculating virtual-water flows we keep track of which parts of these flows refer to green, blue, and gray water.

We have taken into account the international trade in cotton products for the complete set of countries from the Personal Computer Trade Analysis System (PC-TAS) of the International Trade Centre (ITC) in Geneva. This covers trade data from 146 reporting countries disaggregated by product and partner countries for the period 1997–2001 (ITC, 2004). For the calculation of international virtual-water flows, all cotton products were considered as reported in ITC's database. This includes the complete set of cotton products from the commodity groups 12, 14, 15, 23, 60, 61, 62, and 63 in PC-TAS. From group 52, only those products with more than 85% of cotton in their composition were considered.

The calculated virtual-water flows between countries in relation to the international trade in cotton products amount to 204 billion m³/yr at a global scale (an average for the period 1997–2001). About 43% of

Table 9.9 Gross virtual-water export from the major cotton-producing countries related to export of cotton products. Period: 1997–2001.

Country	Virtual-water export (10^9 m^3/yr)			
	Blue	Green	Gray	Total
Argentina	0.85	1.98	0.13	2.95
Australia	2.34	1.44	0.55	4.34
Brazil	0.07	1.03	0.17	1.27
China	9.32	11.36	5.43	26.11
Egypt	1.72	—	0.13	1.85
Greece	1.41	0.41	0.36	2.18
India	5.75	16.83	3.08	25.66
Mali	0.46	1.17	0.11	1.73
Mexico	2.23	1.04	0.86	4.13
Pakistan	10.64	2.87	3.05	16.56
Syria	1.63	0.04	0.07	1.75
Turkey	4.08	0.40	0.89	5.37
Turkmenistan	1.41	0.10	0.31	1.83
Uzbekistan	7.74	0.15	1.66	9.55
USA	4.34	11.18	5.18	20.70
Others	32.73	31.06	13.83	77.62
Global total	86.72	81.05	35.83	203.6

this total flow refers to blue water, about 40% to green water, and about 17% to gray water (Tables 9.9 and 9.10). As we have seen in Chapter 3, the virtual-water flows in relation to international trade in all crop, livestock, and industrial products total 1,625 billion m³/yr at a global scale. The global total of annual gross virtual-water flows between nations related to cotton trade is thus 12% of the total international virtual-water flows.

The countries producing more than 90% of seed cotton are responsible for only 62% of the global virtual-water exports (Table 9.9). This can be understood from the fact that the countries that import raw cotton from the major producing countries export significant volumes again to other countries, often in some processed form. Export of cotton products made from imported raw cotton is significant for instance in Japan, the European Union, and Canada.

Pakistan, China, Uzbekistan, and India are the largest exporters of blue water. These countries export a good deal of water not only in an

Table 9.10   Gross virtual-water imports due to the import of cotton products.
Period: 1997–2001.

Country	Virtual-water import (10^9 m³/yr)			
	Blue	Green	Gray	Total
Brazil	1.5	2	0.4	3.9
Canada	1	1.6	0.6	3.2
China	15.9	15.6	6.7	38.2
France	3.2	2.4	1.2	6.8
Germany	5	3.5	1.8	10.4
Indonesia	2	1.9	0.7	4.6
Italy	4.5	2.9	1.3	8.7
Japan	3.3	3.3	1.5	8.2
Korea Republic	2.8	2.6	1	6.4
Mexico	2.9	6.4	3.2	12.5
Netherlands	1.6	1.4	0.7	3.7
Russian Federation	2.5	0.5	0.6	3.7
Thailand	1.4	1.5	0.5	3.3
Turkey	2.6	1.4	0.7	4.7
UK	3.1	2.9	1.3	7.3
USA	12.2	10	5.3	27.5
Others	21.1	21.2	8.3	50.6
Global total	86.72	81.05	35.83	203.6

absolute sense, but also in a relative sense: more than half of the blue water used for cotton irrigation enters export products. The USA ranks high as a virtual-water exporter due to its large share of green water export. Countries with a large amount of pollution related to the production of export-cotton are China, the USA, and Pakistan.

Water Footprints Related to Consumption of Cotton Products

In assessing a national water footprint due to domestic cotton consumption we distinguish between the internal and the external footprint. The internal water footprint is defined as the use of domestic water resources to produce cotton products consumed by inhabitants of the country. It is the sum of the total volume of water used from the domestic water resources to produce cotton products minus the total volume of virtual-water export related to export of domestically produced cotton products. The external water footprint is defined as the annual volume of water resources used in other countries to produce cotton products consumed by the inhabitants of the country concerned. The external water footprint is calculated by taking the total virtual-water import into the country and subtracting the volume of virtual water exported to other countries as a result of re-export of imported products.

The global water footprint related to the consumption of cotton products is estimated at 256 billion m^3/yr, which is 43 m^3/yr per capita on average. About 42% of this footprint is due to the use of blue water, another 39% to the use of green water, and about 19% to gray water production (Table 9.11). *About 44% of the global water use for cotton growth and processing is not for serving the domestic market but for export.* If we consider not only the water requirements for cotton products but take into account the water needs for the full spectrum of consumed goods and services, the global water footprint is 7,450 billion m^3/yr (see Chapter 5). This includes the use of green and blue water but it largely excludes the water requirement for dilution of waste flows. As a proxy for the latter we take here the rough figure provided by Postel et al. (1996), who estimate the global dilution water requirement at 2,350 billion m^3/yr. This means that the full global water footprint is about 9,800 billion m^3/yr. The global water footprint related to cotton consumption is 256 billion m^3/yr,

Table 9.11 The global water footprint due to cotton consumption. Period: 1997–2001.

	Global water footprint (10^9 m^3/yr)			
	Blue	Green	Gray	Total
Internal*	59.6	54.8	28.5	143
External*	48.0	44.7	20.7	113
Total	108	99	49	256
Percentage	42%	39%	19%	—

* The internal water footprint at a global scale refers to the aggregated internal water footprints of all nations of the world. The external water footprint refers here to the aggregated external water footprints of all nations.

which means that the consumption of cotton products accounts for 2.6% of the full global water footprint.

The countries with the largest impact on foreign water resources are China, the USA, Mexico, Germany, the UK, France, and Japan (Table 9.12). About half of China's water footprint due to cotton consumption is within China (the internal water footprint); the other half (the external footprint) impacts on other countries, mainly India (predominantly green water use) and Pakistan (predominantly blue water use).

For each country, the water footprint as a result of domestic cotton consumption can be mapped as has been done for the USA in Map 13. The arrows show the tele-connections between the area of consumption (the USA) and the areas of impact (notably India, Pakistan, China, Mexico, and the Dominican Republic). The total water footprint of an average US citizen due to the consumption of cotton products is $135 \, m^3/yr$ – more than three times the global average – out of which about half is from the use of external water resources. If all world citizens were to consume cotton products at the US rate, other factors remaining equal, global water use would increase by 5% (from 9,800 to 10,300 billion m^3/yr), which is quite substantial given that the world's population already uses more than half of the runoff water that is reasonably accessible (Postel et al., 1996).

For proper understanding of the impacts shown in Map 13 it should be observed here that the map shows the full internal water footprint of the USA plus the external water footprints in other countries insofar as easily traceable. For instance, the USA imports several types of cotton products from the EU, that together contain 430 million m^3/yr of virtual water, but these cotton products do not fully originate from the 25 countries of the European Union (EU25). In fact, the EU25 imports raw cotton, gray fabrics, and final products from countries such as India, Uzbekistan, and Pakistan, then partly or fully processes these products into final products and ultimately exports to the USA. Out of the 430 million m^3/yr of virtual water exported from the EU25 to the USA, only 16% is water appropriated within the EU25; the other 84% refers to water use in countries from which the EU25 imports (e.g. India, Uzbekistan, and Pakistan). For simplicity, we show in the map only the "direct" external footprints (tracing the origin of imported products only one step back), and not the "indirect" external footprints. Adding the latter would mean

Table 9.12 The water footprint of cotton consumption for selected countries. Period: 1997–2001.

Country	Internal water footprint (10^6 m³/yr)			External water footprint (10^6 m³/yr)			Total (10^6 m³/yr)
	Blue	Green	Gray	Blue	Green	Gray	
Argentina	832	1,953	156	22	89	20	3,071
Australia	755	585	296	234	294	164	2,328
Belgium–Luxembourg	15	0	25	1,215	763	395	2,414
Brazil	404	3,454	804	1,451	1,643	369	8,126
Canada	39	0	86	592	1,204	478	2,399
China	8,775	11,176	6,585	10,738	10,213	4,485	51,972
Colombia	174	160	115	170	357	98	1,074
Egypt	1,433	0	177	60	193	25	1,888
France	53	0	93	2,387	1,576	867	4,977
Germany	47	0	79	3,525	2,049	1,220	6,920
Greece	1,199	416	382	278	266	115	2,657
India	7,015	19,462	3,965	281	222	81	31,024
Indonesia	86	18	152	773	683	330	2,042
Iran	789	731	353	32	4	7	1,917
Israel	124	124	72	452	814	241	1,828
Italy	83	0	106	2,254	644	465	3,552
Japan	78	0	165	1,696	1,735	935	4,610
Korea Republic	124	0	224	1,808	1,538	648	4,343
Malaysia	36	0	68	609	686	262	1,662

(Continued)

Table 9.12 (*Continued*).

Country	Internal water footprint (10^6 m^3/yr)			External water footprint (10^6 m^3/yr)			Total (10^6 m^3/yr)
	Blue	Green	Gray	Blue	Green	Gray	
Mexico	460	327	549	1,297	5,395	2,489	10,517
Netherlands	22	0	39	1,277	1,035	539	2,912
Nigeria	658	613	311	93	200	48	1,924
Pakistan	9,672	2,567	3,012	0	0	0	15,251
Poland	34	0	55	769	274	215	1,347
Russian Federation	84	0	143	2,076	74	496	2,874
Singapore	17	0	31	708	857	361	1,974
Spain	387	325	173	693	518	232	2,328
Syria	1,736	45	166	0	0	0	1,947
Thailand	106	42	136	690	766	243	1,984
Turkey	3,754	508	1,172	1,453	1,106	482	8,476
Turkmenistan	3,958	287	897	1	0	0	5,143
UK	35	0	62	2,307	2,175	980	5,560
USA	5,111	9,314	4,971	9,429	5,738	3,216	37,780
Uzbekistan	6,956	131	1,598	0	0	0	8,685
Global total	59,605	54,793	28,515	48,025	44,655	20,743	2,56,336

adding for instance an arrow from India to the EU25, which is then forwarded to the USA. Doing so for all indirect external water footprints would create an incomprehensible map. For the same reason, we have shown arrows for only the largest virtual-water flows toward the USA.

The water footprint as a result of cotton consumption in Japan is mapped in Map 14. For their cotton the Japanese consumers rely most importantly on the water resources of China, Pakistan, India, Australia, and the USA. Japan does not grow cotton, and also does not have a large cotton processing industry. The Japanese water footprint due to consumption of cotton products is 4.6 billion m^3/yr, of which 95% impacts on other countries. The cotton products imported from Pakistan put considerable pressure on Pakistan's scarce blue water resources. In China and even more so in India, cotton is produced with lower inputs of blue water (in relation to green water), so that cotton products from China and India put less stress per unit of cotton product on scarce blue water resources than in Pakistan.

Map 15 shows the water footprint due to cotton consumption in the EU25. Some 84% of the EU's cotton-related water footprint lies outside the EU. From the map it can be seen that, for their cotton supply, the European community depends most heavily on the water resources of India. This puts stress on the water availability for other purposes in India. In India one third of the cotton harvest area is irrigated; cotton imports from these irrigated areas have a particularly large opportunity cost, because the competition for blue water resources is higher than for green water resources. If we look at the impacts of European cotton consumption on blue water resources, these are even higher in Uzbekistan than in India. Uzbekistan uses 14.6 billion m^3/yr of blue water to irrigate cotton fields, of which it exports 3.0 billion m^3/yr in virtual form to the EU25. The consumers in the EU25 countries thus indirectly (and mostly unconsciously) contribute about 20% toward the desiccation of the Aral Sea. In terms of pollution, cotton consumption in the EU25 has largest impacts in India, Uzbekistan, Pakistan, Turkey, and China. These impacts are due partly to the use of fertilizer in the cotton fields and partly to the use of chemicals in the cotton processing industries. Cotton consumption in the EU25 also causes pollution in the region itself, mainly from the processing of imported raw cotton or gray fabrics into final products.

The three components of a water footprint – green water use, blue water use, and gray water production – affect water systems in different ways. Use of blue water generally affects the environment more than green water use. Blue water is lost to the atmosphere whereas otherwise it would have stayed in the ground or river system from which it was taken. Green water on the other hand would have been evaporated through another crop or through natural vegetation if it had not been used for cotton growth. Therefore there should generally be more concern with the blue than with the green water footprint. The gray water footprint deserves attention as well, since pollution is a choice and not a necessity. Waste flows from cotton industries can be treated so that no dilution water would be required at all. An alternative to treatment of waste flows is reduction of waste flows. With cleaner production technology, the use of chemicals in cotton industries can be reduced by 30%, with a reduction in the COD content in the effluent of 60% (Visvanathan et al., 2000).

Sustainable Use of Water

We believe that a single indicator of sustainability does not exist, because of the variety of facts, values, and uncertainties that play a role in any debate on sustainable development. The water footprint of a nation should clearly not be seen as the ultimate indicator of sustainability, but rather as a new indicator that can add to the sustainability debate. It adds to the ecological footprint and the embodied energy concept by taking water as a central viewpoint as an alternative to land or energy. It adds to earlier indicators of water use by taking the consumer's perspective on water use instead of the producer's perspective.

Following the introduction of the ecological-footprint concept in the 1990s, several scholars have expressed doubts about whether the concept is useful in science or policy making. At the same time we see that the concept attracts attention and evokes scientific debate. We expect that the water-footprint concept will lead to a similar dual response. On the one hand the water footprint simply gathers and presents known data in a new format and as such does not add new knowledge. On the other hand, the water footprint adds a fruitful new perspective on issues such as water scarcity, water dependency, sustainable water use, and the implications of global trade for water management.

For water managers, water management is a river basin or catchment issue (see for instance the new South African National Water Act, 1998, and the new European Water Framework Directive, 2000). The water footprint, showing the use of water in foreign countries, demonstrates that it is not sufficient to stick to that scale. Water problems in the major cotton-producing areas of the world cannot be solved without addressing the global issue that consumers are not being held responsible for some of the economic costs and ecological impacts of their cotton consumption, which remain in the producing areas. The water footprint shows water use from the consumer's perspective, while traditional statistics show water use from the producer's perspective. This makes it possible to compare the water demand for North American or European citizens with the water demand for people in Africa, India, or China. In the context of equitability and sustainability, this is a more useful comparison than one between the actual water use in the USA or Europe and the actual water use in an African or Asian country, simply because the actual water use tells us about production but nothing about consumption.

The water footprint shows how dependent many nations are on the water resources in other countries. For its consumption of cotton products, the EU25 is hugely dependent on the water resources in other continents, particularly Asia as we have shown, but there is a strong dependence on foreign water resources for other products also. This means that water in Europe is scarcer than current indicators (showing water abstractions within Europe in relation to the available water resources within Europe) suggest.

Cotton consumption is responsible for 2.6% of global water use. As a global average, 44% of the water use for cotton growth and processing is not for serving the domestic market but for export. This means that – roughly speaking – nearly half of the water problems in the world related to cotton growth and processing can be attributed to foreign demand for cotton products. By looking at the trade relations, it is possible to track down the location of the water footprint of a community or, put another way, to link consumption in one place to the impacts elsewhere. Our calculations show for instance that consumers in the EU25 countries indirectly contribute about 20% toward the desiccation of the Aral Sea. Visualizing the actual but hidden link between cotton consumers and the water impacts of cotton production is a relevant issue in light of the fact that the economic and environmental impacts of water use are generally not included in the price paid by the foreign consumers for the cotton products. Including information about the water footprint with the product, be it in the form of pricing or product labeling, is thus a crucial aspect in policy aimed at the reduction of negative impacts such as water depletion and pollution. Given the global character of the cotton market, international cooperation in setting the rules for cotton trade is a precondition.

Since each component of the total water footprint includes a certain economic cost and environmental impact, it would be useful to see which of the costs and impacts are transferred to the consumer. We have not done a careful examination of this, but there is considerable evidence that the majority of costs and impacts of water use and pollution caused by agriculture and industry are not translated into the price of products. According to the World Bank, the economic cost recovery in developing countries in the water sector is about 25% (Serageldin, 1995). Social and environmental impacts of water use are generally not translated into the price of products at all, with the exception sometimes of the costs of wastewater treatment before disposal. Most of the global waste flows are not treated however. Although a few industrialized countries achieve a wastewater-treatment coverage of nearly 100%, this coverage remains below 5% in most developing countries (Hoekstra, 1998; EC Statistical Office, 2006). Besides, the 100% treatment coverage in some of the industrialized countries refers to treatment of concentrated waste flows from households and industry only, but excludes the diffuse waste flow from agriculture. Given the general lack of proper water pricing mechanisms or other ways of transmitting production-information, cotton consumers have little incentive to take responsibility for the impacts on remote water systems.

About one fifth of the global water footprint due to cotton consumption is related to pollution. This estimate is based on the assumption that wastewater flows can be translated into a certain water requirement for dilution based on water quality standards. Implicitly we have assumed here that the majority of waste flows enter natural water bodies without prior treatment, which is certainly true for leaching of fertilizers in agriculture and largely true for waste flows from cotton industries. In some rich countries, however, waste flows from industries are often treated before disposal, meaning we will have overestimated dilution water requirements. In the case of treatment of waste flows to the extent that the effluents meet water quality standards, a better estimate for the water requirement would be to consider the actual water use for the treatment process. Another issue is that we did not account for natural background concentrations in dilution water, meaning that our estimate for the required dilution volume is conservative. The fact that we looked at the dilution volume required for fertilizers but not at the volume for diluting pesticides used will also make our estimate conservative.

Chapter 10

Water as a Geopolitical Resource

Water is a basic natural resource, indispensable for life. Although not priced, the value of rain is undisputed. Income and food supply of millions of people are directly dependent on the availability of freshwater. In this sense, many people are heavily dependent on water. In this chapter, however, we will not expand on this issue of individual people depending on water for their livelihood. We will not discuss how farmers depend on water, but instead how whole nations can be water dependent.

Nations can be "water dependent" in two different ways. They can be dependent on water that flows in from neighboring countries and they can be dependent on virtual-water import. An example of the first is Egypt depending on the water of the Nile. This type of water dependency occurs when the external water resources of a country constitute a significant part of the total renewable water resources of the country. FAO (2003a) defines the "external renewable water resources" of a country as the part of the country's renewable water resources that is not generated in the country. It consists of the inflows from upstream countries (groundwater and surface water) and part of the water of border lakes or rivers. A distinction is made between the "natural" and the "actual" external renewable water resources. The first term refers to the natural incoming flow originating outside the country; the actual external resources are usually less than the natural external resources, because in this case upstream water abstractions are subtracted, as are water flows reserved for upstream and downstream countries through formal or informal agreements or treaties. The "internal renewable water resources" of

Table 10.1 Dependency on incoming river flows for selected countries.

Country	Internal renewable water resources* (10^9 m^3/yr)	External (actual) renewable water resources* (10^9 m^3/yr)	External water resources dependency[†] (%)
Iraq	35	40	53
Cambodia	121	356	75
Pakistan	52	170	77
Netherlands	1.1	80	88
Egypt	1.8	56.5	97

* *Source:* FAO (2003a, 2006a).
[†] Defined as the ratio of the external to the total renewable water resources.

a country concern the average annual flow of rivers and recharge of aquifers generated by endogenous precipitation. The total renewable water resources of a country are the sum of internal and external renewable water resources. Table 10.1 shows the "external water resources dependency" for some selected downstream countries. For a country like Egypt the dependency is extremely high, because the country receives hardly any precipitation and thus depends mostly on the inflowing Nile water. Similarly, but to a lesser extent, Pakistan depends strongly on the water of the Indus, Cambodia on the water of the Mekong, and Iraq on the Tigris and Euphrates. In all these cases water is an important geopolitical resource, affecting power relations between the countries that share a common river basin. In a country like the Netherlands external water resources dependency is high but less important, because water is less scarce than in the previous cases. Nevertheless, here too there is a dependency, since activities within the upstream countries undoubtedly affect downstream low flows, peak flows, and water quality.

The political relevance of "external water resources dependency" of nations makes water a *regional* geopolitical resource in some river basins. There is a huge amount of literature on conflicts and cooperation in transboundary rivers (e.g. Green Cross International, 2000; Bogardi and Castelein, 2003). In this book we have not addressed this subject. This book is about the other type of water dependency, namely virtual-water import dependency, which makes water a *global* geopolitical resource. The reason that water is of global political interest lies in the combination of a number of factors: water is a vital natural resource, its unique character prevents substitution by other resources, its uneven distribution throughout the world creates haves and have-nots, and its scarcity has been continually increasing during the past decades. Where in the past water-abundant regions did not fully exploit their potential, they now increasingly do so by exporting water in virtual form or even in real form. The other side of the coin is the increasing dependency of water-scarce nations on the supply of food or water, which can be exploited politically by those nations which control the water.

From a water resources point of view one might expect a positive relation between water scarcity and virtual-water import dependency, particularly in the areas of great water scarcity. Water scarcity can be defined as a country's total water footprint divided by the country's water availability (see Appendix I). Virtual-water import dependency can be defined as the ratio of the external water footprint of a country to its total water footprint. Countries with a very high degree of water scarcity – Kuwait, Qatar, Saudi Arabia, Bahrain, Jordan, Israel, Oman, Lebanon, and Malta, for example – indeed have a very high virtual-water import dependency (> 50%). The water footprints of these countries have been largely externalized. Jordan annually imports a virtual-water quantity five times its own yearly renewable water resources. Although saving its domestic water resources, it makes Jordan heavily dependent on other nations, for instance the USA. Other water-scarce countries with high virtual-water import dependency (25–50%) include Greece, Italy, Portugal, Spain, Algeria, Libya, Yemen, and Mexico. Even European countries that do not have an image of being water scarce, such as the UK, Belgium, the Netherlands, Germany, Switzerland, and Denmark, have a high virtual-water import dependency. Table 10.2 presents the data for some selected countries. Data for all countries are included in Appendix V. The average virtual-water import dependency is 16%. With continued global trade liberalization, international water dependencies are likely to increase.

In most water-scarce countries the choice is either (over)exploitation of the domestic water resources in order to increase water self-sufficiency (the apparent strategy of Egypt) or virtual-water import at the cost of becoming water dependent (e.g. Jordan). For some countries, the choice is hardly a choice at all. In Yemen, for example, water

Table 10.2 Virtual-water import dependency of selected countries. Period: 1997–2001.

Country	Internal water footprint (10^9 m^3/yr)	External water footprint (10^9 m^3/yr)	Water self-sufficiency* (%)	Virtual-water import dependency[†] (%)
Indonesia	242	28	90	10
Egypt	56	13	81	19
South Africa	31	9	78	22
Mexico	98	42	70	30
Spain	60	34	64	36
Italy	66	69	49	51
Germany	60	67	47	53
Japan	52	94	36	64
UK	22	51	30	70
Jordan	1.7	4.6	27	73
Netherlands	4	16	18	82

* Defined as the ratio of the internal to the total water footprint.
† Defined as the ratio of the external to the total water footprint.

is very scarce and groundwater resources are heavily overdrawn, but virtual-water imports are limited for the simple reason that the country does not have the foreign currency to import water-intensive commodities in order to prevent overexploitation of its domestic water resources.

The two largest countries in the world, China and India, still have a very high degree of national water self-sufficiency (93% and 98%, respectively). However, the two countries currently have relatively low water footprints per capita (China 700 m^3/yr and India 980 m^3/yr). If the consumption pattern in these countries changes to that of the USA or some Western European countries, they will be facing severe water scarcity in the future and will probably be unable to sustain their high degree of water self-sufficiency. A relevant question is how China and India are going to feed themselves in the future (Brown, 1995; WWI, 2006). If they were to decide to obtain food security partly through food imports, this would put enormous demands on the land and water resources in the rest of the world.

Historically, water has been seen as a local, not a global, resource. When talking about oil, people have no problem in accepting the geopolitical implications of its uneven distribution, but the message that freshwater is a geopolitical resource of global importance as well does not seem to get across to many people. Oil and freshwater, however, share two important characteristics: they are both important production factors and they are both unevenly distributed across the globe. Water is in fact a sort of "white oil" or "blue gold" as others have put it (Donkers, 1994; Barlow and Clarke, 2002). It would be better to acknowledge this fact. It would put water issues higher on the international political agenda, where they deserve to be.

Chapter 11

Efficient, Sustainable, and Equitable Water Use in a Globalized World

Globalization of freshwater brings both opportunities and risks. The most obvious opportunity of reduced trade barriers is that virtual water can be regarded as a possibly cheap alternative source of water in areas where freshwater is relatively scarce. Virtual-water import can be used by national governments as a tool to release the pressure on their domestic water resources. In an open world economy, according to international trade theory, the people of a nation will seek profit by trading products that are produced with resources that are (relatively) abundant within the country for products that need resources that are (relatively) scarce. People in countries where water is a very scarce resource could thus aim to import products that require a lot of water (water-intensive products) and export products or services that require less water (water-extensive products). This import of virtual water (as opposed to real water, which is generally too expensive) will relieve the pressure on the nation's own water resources. For water-abundant countries an argument can be made for export of virtual water. Trade can physically save water if products are traded from countries with high to countries with low water productivity. For example, Mexico imports wheat, maize, and sorghum from the USA, which requires 7.1 billion m^3 of water per year in the USA. If Mexico were to produce the imported crops domestically, it would require 15.6 billion m^3 of water per year. Thus, from a global perspective, the trade in cereals from the USA to Mexico saves 8.5 billion m^3/yr. Although there are also examples where water-intensive commodities flow in the opposite direction, from countries with low to countries with high water productivity, various studies indicate that the resultant of all international trade flows works in a positive direction (De Fraiture et al., 2004; Oki and Kanae, 2004; Chapagain et al., 2006a; Yang et al., 2006). In Chapter 4 we showed that international trade in agricultural commodities reduces global water use in agriculture by 5%. Liberalization of trade seems to offer new opportunities to contribute to a further increase of efficiency in the use of the world's water resources.

A serious drawback of trade is that the indirect effects of consumption are externalized to other countries. While water in agriculture is still priced far below its real cost in most countries, an increasing volume of water is used for processing export products. The costs associated with water use in the exporting country are not included in the price of the products consumed in the importing country. Consumers are generally not aware of – and do not pay for – the water problems in the overseas countries where their goods are being produced. According to economic theory, a precondition for trade to be efficient and fair is that consumers bear the full cost of production and impacts.

Another downside of intensive international virtual-water transfers is that many countries increasingly depend on the import of water-intensive commodities from other countries. As we saw in Chapter 10, Jordan annually imports a virtual-water volume that is five times its own annual renewable water resources. Other countries in the Middle East, but also various European countries, have a similar high water import dependency. The increasing lack of self-sufficiency has made various individual countries, but also larger regions, very vulnerable. If for whatever reason food supplies cease – be it due to war or a natural disaster in an important export region – the importing regions will suffer severely. A key question is to what extent nations are willing to take this risk. The risk can be avoided only by promoting national self-sufficiency in water and food supply (as Egypt and China do). The risk can be reduced by importing food from a wide range of trade partners. The current worldwide trend, however, facilitated by the World Trade Organization, is toward reducing trade barriers and encouraging free international trade, and decreasing interference by national governments.

Fairness and Sustainability of Large Water Footprints

Two other issues in the context of globalization are the equitable and sustainable use of the world's natural resources. Some people around the world have comparatively large water footprints, while others have small ones. This raises the question of whether this is fair and sustainable. Under current production conditions it would be impossible for all world citizens to develop a water footprint the same size as the present water footprint of the average US citizen. People in the USA have, on average, the largest water footprint per capita in the world, namely 2,480 m^3/yr. China has an average water footprint of 700 m^3/yr per capita, while the world average is 1,240 m^3/yr (see Chapter 5). The issues of fairness and sustainability become very obvious in this imaginary growth scenario, but both are already relevant today.

Currently, more than 1 billion people do not have access to clean drinking water (UNESCO, 2006), while others water their gardens, wash their cars, fill their swimming pools, and enjoy the availability of water for many other luxury purposes. In addition, many people consume a large amount of meat, which significantly increases their water footprint. The average meat consumption in the USA for instance is 120 kg/yr, more than three times the world-average. The water used to produce the feed for the animals that provide the meat for the rich cannot be used for other purposes, for example to fulfill more basic needs of people who however cannot afford to pay. The question of whether the current distribution of water footprints is fair is a political one. Redistribution of welfare among individuals is normally done within the borders of the nation state, but since the distribution of water and water-intensive products is very uneven across the globe, the redistributive issue becomes a global matter as well. The normative question at global level is whether wealthy water-rich nations should play a role in supporting developing water-poor nations, for instance by helping them to efficiently and sustainably use their scarce water resources.

What is a "sustainable water footprint," given the 6 billion inhabitants on earth and the fact that the total water availability in the world is limited? The current global water footprint is 7,450 billion m^3/yr, which in many places obviously leads to unsustainable conditions, as witnessed by the many reported cases of water depletion and pollution (UNESCO, 2003, 2006). Although the annual volume of precipitation over land is roughly known, it is very difficult to give a global figure for the maximum "sustainable water footprint" as an upper limit to global water use. There are various reasons for this. One is that not all precipitation can be used productively, because its fall is unevenly spread in time and space, so that there are places and times when the water will inevitably flow to the oceans. According to Postel et al. (1996) about 20% of total runoff forms remote flows that cannot be appropriated and 50% forms uncaptured floodwater, so that only 30% of runoff remains for use. Although there has been some research in this direction, it is not yet clearly established what fraction of this remaining flow should remain untouched in order to fulfill the environmental flow requirements (Smakhtin et al., 2004). It has also not been established what fraction of the total evapotranspiration on land may be counted as potentially productive. Finally, what we would consider the maximum "sustainable water footprint" at global level depends on what assumptions are made with respect to the level of technology. One could take water productivities as they are in practice at present (which differ from location to location), or one could work with the potential water productivities based on existing technology. The latter would lead to a more optimistic figure than the former, but also a less realistic one. So far no estimates of the world's maximum "sustainable water footprint" have been made, but a general feeling exists that if it has not passed it already, the current global water footprint will not be far below the maximum sustainable value – witness the widely promoted need for water demand management and water use efficiency improvements (Postel et al., 1996; FAO, 2003c; UNESCO, 2003, 2006). This brings us back to the issue of fairness: Is it fair if some people use more than an equitable share of the maximum global volume of annually available water resources? The average person in North America and Southern Europe certainly does so.

Global Rules of the Game

In order to benefit from the opportunities and to avert the downsides of international trade, nations can develop their own strategies. Measures to redistribute wealth and promote sustainability can also be taken at national level. However, in the end, efficient and fair international trade, equitability among people, and sustainable use of the world's water resources are true global issues that are likely to benefit from shared "rules of the game." In the remainder of this book we explore what sorts of rules ("institutional arrangements," in the jargon of policy science) could be employed in order to promote efficient, sustainable, and equitable use of water resources in the world. We will identify a few possible types of arrangements in an explorative manner. By "explorative" we mean that we do not intend to be exhaustive and that identification of possible types of arrangements will have priority over reviewing the political feasibility of the arrangements identified. In fact, at this stage we should not care too much about what seems feasible and what looks unlikely. We deal with issues that have not been addressed before, so the tools will not be the same as any seen previously. Our view is that some speculation on what would be required – and on some possible mechanisms – without the hindrance of day-to-day politics might be more productive than thinking within existing structures.

An International Protocol on Water Pricing

First of all, there is a need to arrive at a global agreement on water pricing structures that cover the full cost of water use, including investment costs, operational and maintenance costs, a water scarcity rent, and the cost of negative external impacts of water use. Without an international treaty on proper water pricing it is unlikely that a globally efficient pattern of water use will ever be achieved. The need to have full-cost pricing has been acknowledged since the Dublin Conference in 1992 (ICWE, 1992). A global ministerial forum to pursue agreements on this does exist in the regular World Water Forums (Morocco 1997, The Hague 2000, Japan 2003, Mexico 2006), but these forums have not been used to take up the challenge of making international agreements on the implementation of the principle that water should be considered as a scarce, economic good. It is not sufficient to leave the implementation of this principle to national governments without having some kind of international protocol on the implementation, because unilateral implementation can be expected to be at the cost of the countries moving ahead. The competitiveness of the producers of water-intensive products in a country that implements a stringent water pricing policy in isolation will be affected, and this, together with the natural resistance of domestic consumers to higher prices of local products, will reduce the feasibility of a unilateral implementation of a rigorous water pricing strategy. An international protocol on full-cost water pricing would contribute to the sustainable use of the world's water resources, because water scarcity would be translated into a scarcity rent and thus affect consumer decisions, even if those consumers live at a great distance from the production site. Such a protocol would also contribute to fairness, by making producers and consumers pay for their contribution to the depletion and pollution of water. Finally, such a protocol would shed fresh light upon the economic feasibility of plans for large-scale inter-basin transfers, since it would force negative external impacts and opportunity costs to be taken into account. Full-cost water pricing should be combined with a minimum water right, in order to prevent poor people being unable to obtain their basic needs.

A Water Label for Water-Intensive Products

A second global arrangement could be a water label for water-intensive products, comparable to the FSC (Forest Stewardship Council) label for wood products. Such a label would make consumers aware of the actual, but so far hidden, link between a consumer product and the impacts on water systems that occur during production. A water label should give a guarantee to the consumer that the product was produced under some clearly defined conditions. The label could be introduced first for a few commodities that usually have great impacts on water systems, such as rice, cotton, paper, and cane sugar. Given the global character of the markets for these goods, international cooperation in setting the labeling criteria and in the practical application of the water label is a precondition. Consideration could be given to integrating the water label within a broader environmental label, but this would probably create new bottlenecks for implementation, so that a first step could be to agree on a separate water label.

Minimum Water Rights

Equitability and sustainability in water use require the establishment of both minimum water rights and maximum levels of water use. The latter has received little attention from the international community and will be discussed further below. The issue of minimum water rights has had more consideration (Gleick, 1998; WHO, 2003; Salman and McInerney-Lankford, 2004). At international level efforts have been made to have access to clean drinking water accepted as a human right. The Universal Declaration of Human Rights from 1948 does not mention access to water as a human right, but the first paragraph of article 25 reads: "Everyone has the right to a standard of living adequate for the health and well-being of himself and of his family, including food, clothing, housing and medical care and necessary social services... ." With a little good will, one could say that the right to a certain minimum of water is thereby implicitly established. A step toward the more explicit formulation of the right to water was made in 1976 with article 12 of the International Covenant on Economic, Social, and Cultural Rights, which acknowledges "the right of everyone to the enjoyment of the highest attainable standard of physical and mental health." In 2000 the Committee on Economic, Social, and Cultural Rights of the United Nations (in her General Comment No. 14) accepted a supplement to this covenant which states that "the right to health embraces a wide range of socio-economic factors that promote conditions in which people can lead a healthy life, and extends to the underlying determinants of health, such as food and nutrition, housing, access to safe and potable water and adequate sanitation, safe and healthy working conditions, and a healthy environment." In 2002 the same committee specified the right to water in her General Comment No. 15: "The human right to water entitles everyone to sufficient, safe, acceptable, physically accessible and affordable water for personal and domestic uses. An adequate amount of safe water is necessary to prevent death from dehydration, to reduce the risk of water-related disease and to provide for consumption, cooking, personal and domestic hygienic requirements."

With these statements the human right to water has been formally established, but there are no enforcement mechanisms. Besides, the right specifically refers to water for basic needs in domestic use, not to water for food. Food itself as a human right had already been established explicitly in article 25 of the Universal Declaration of Human Rights. Although one cannot deny that the right to food translates into a certain volume of water required to produce the food, the right

to food has never been translated into a "right to water for food."
On the level of the individual this is also not useful, because it would
wrongly presuppose that every individual produces his or her own
food. However, the right to food implies that every individual has a
sort of "claim" on a certain volume of the world's water resources
that is required to produce the amount of food that he or she is
entitled to according to the existing right to food. Given the uneven
distribution of water across the world, an important question is: How
do the existing human rights to water and food translate into a moral
obligation of communities that have abundant water resources
at their disposal toward communities with severely limited water
resources? One of the concrete steps taken by the international community has been the formulation of the Millennium Development
Goals during the UN Millennium Summit in New York in 2000.
Definite targets include to reduce by half the proportion of people
who suffer from hunger and also to reduce by half the proportion
of people without sustainable access to safe drinking water (both
targets referring to the period 1990–2015). The weak point of the
Millennium Development Goals is that they lack a clear course of
action and a mechanism for enforcement. As a result, there is no
guarantee that the good intentions will be realized.

(Tradable) Water-Footprint Permits

The issues of fair water allocation and sustainable water use cannot
be solved by minimum water rights alone, but also require arrangements about maximum levels of water use. The limited availability of
freshwater in the world puts a maximum on the human global water
footprint. The question for the global community is how this global
maximum can be transferred to the national or even the individual
level. Or in other words: What is each nation's and each individual's
"reasonable" share of the globe's water resources? And what mechanisms could be established in order to ensure that people do not use
more than their "reasonable" share? Even in the case of full-cost water
pricing, there is no guarantee that the globe's water resources will
be used in a sustainable manner. The reason is that proper marginal-
cost pricing throughout the economy will lead to the so-called "economic optimum" (provided that some other basic presuppositions

of economic theory are met as well), but this economic optimum is not necessarily sustainable. Consumers, even if they pay full marginal costs, do not necessarily make choices that can be maintained in the long term. Economists in particular tend to forget this. Maximum levels of water use to guarantee sustainability could be formalized in the form of a water-footprint permit system. An international protocol on establishing water-footprint permits would be comparable to the Kyoto Protocol on the emission of greenhouse gases (drafted in 1997, effective since 2005). The Kyoto Protocol is based on the understanding that, to prevent human-induced climate change, there is a ceiling on the maximum volume of greenhouse gas emissions from human activities that can be accommodated by the global system. The fact that it is not known exactly what this ceiling is has apparently not held the international community back in setting political targets with respect to greenhouse gas emission reductions. The same would have to happen if the international community were willing to set targets with respect to maximum water footprints, because the precise ceiling on water use is subject to debate as well, as explained earlier. In the case of the Kyoto Protocol, the maximum allowable emission permits have been issued in the form of tradable emission permits. In the case of a protocol on water use, this could be done in the form of tradable water-footprint permits.

Global Arrangements versus the Subsidiarity Principle

The above arguments for coordination at global level seem to be at odds with the subsidiarity principle, nowadays widely accepted and promoted in the field of water governance. This principle means that water issues should be settled at the lowest community level possible. As has been argued in this book, however, some water issues have a truly global character and cannot be solved at a lower community level than that of the global community. Hence, strictly speaking, there is no conflict with the subsidiarity principle. However, it is a fact that global arrangements in the area of water governance will definitely subtract from the mandates at lower community levels. Finding a balance between formal arrangements at different levels of governance will indeed be a true challenge.

Globalization: Pro or Anti?

With increasing globalization of trade, there are opportunities to enhance global water use efficiency if nations make use of their comparative advantages. At the same time, however, national water dependencies and overseas external impacts are likely to increase. In addition, since freshwater is gradually becoming a global resource (demand and supply match at the global level rather than at the river basin level), equitable and sustainable water use are turning into global issues as well. We think that global arrangements based on ideas such as a global water pricing protocol, water labeling, minimum water rights, and a water-footprint permit system are necessary to promote water use efficiency and at the same time ensure sustainable water use and encourage equitable sharing of the limited water resources of the world. In the heated debate about globalization we take the stance of neither pro-globalist nor anti-globalist. What is really needed in our view is the establishment of proper arrangements at the global level where national arrangements are not sufficient.

Appendix I

Analytical Framework for the Assessment of Virtual-Water Content, Virtual-Water Flows, Water Savings, Water Footprints, and Water Dependencies

During the past 5 years we have developed a comprehensive set of definitions and calculation methods for assessing the virtual-water content of products, virtual-water flows between nations, water footprints of nations, and various other parameters. We list details of them here. Where we use terms that are common in the literature (e.g. crop water requirement, effective rainfall, irrigation requirement, yield) we use the same definitions as the Food and Agriculture Organization (FAO, 2006a,b,e). For other terms we use the definitions as given in the Glossary.

Virtual-Water Content of Products

The virtual-water content of a product is the volume of freshwater used to produce the product, which depends on the water use in the various steps of the production chain. The virtual-water content of a product breaks down into green, blue, and gray components. These components refer to evaporated rainwater, evaporated ground-/surface water, and polluted water, respectively. For primary products (e.g. wheat), there is only one production step to be considered, but for processed products (e.g. beef) there are more steps. Below we present first our method to calculate the virtual-water content of primary crops, i.e. crops as they come directly from the land without having undergone any real processing. Second, we present our method to calculate the virtual-water content of crop products (processed from primary crops). Third, we show how to calculate the virtual-water content of live animals and livestock products. Finally, we show how we have crudely assessed the virtual-water content of industrial products and the virtual-water content of domestic water supply.

The virtual-water content of a primary crop can be calculated in a number of steps. First, crop water requirements (CWR, mm/day) have to be calculated over the period from planting to harvest. The crop water requirement is the water needed for evapotranspiration under ideal growth conditions, measured from planting to harvest. "Ideal conditions" means that adequate soil water is maintained by rainfall and/or irrigation, so that water does not limit plant growth and crop yield. The crop water requirements of a certain crop under particular climatic conditions can be estimated with the models developed by the Food and Agriculture Organization (Allen et al., 1998; FAO, 2006b). Basically, the crop water requirement is calculated by multiplying the reference crop evapotranspiration (ET_0, mm/day) by the crop coefficient (K_c):

$$CWR = K_c \times ET_0$$

The reference crop evapotranspiration is the evapotranspiration rate from a reference surface, not short of water. The reference is a hypothetical surface with extensive green grass cover with specific characteristics. The only factors affecting ET_0 are climatic parameters. ET_0 expresses the evaporating power of the atmosphere at a specific location and time of the year and does not consider the crop characteristics and soil factors. The actual crop evapotranspiration under ideal conditions differs markedly from the reference crop evapotranspiration, as the ground cover, canopy properties, and aerodynamic resistance of the crop are different from grass. The effects of characteristics that distinguish field crops from grass are integrated into the crop coefficient (K_c).

After crop water requirements, irrigation requirements have to be calculated over the full growing period. The irrigation requirement (IR, mm/day) is zero if effective rainfall is greater than the crop water requirement at a certain point in time, but otherwise it is equal to the difference between the crop water requirement (CWR, mm/day) and effective rainfall (P_{eff}, mm/day):

$$IR = \max(0, CWR - P_{eff})$$

Effective rainfall is that part of the total amount of rainwater useful for meeting the water need of the crop – generally slightly less than the total rainfall because not all rainfall can actually be appropriated by the crop, for example due to surface runoff or percolation (Dastane, 1978).

Green water evapotranspiration (ET_g, mm/day), i.e. evapotranspiration of rainfall, will be equal to the minimum of crop water requirement (CWR, mm/day) and effective rainfall (P_{eff}, mm/day). Similarly, blue water evapotranspiration (ET_b, mm/day), i.e. field-evapotranspiration of irrigation water, will be the minimum of irrigation requirement (IR, mm/day) and effective irrigation (I_{eff}, mm/day):

$$ET_g = \min\left(CWR, P_{eff}\right)$$

$$ET_b = \min\left(IR, I_{eff}\right)$$

Effective irrigation refers to the amount of irrigation water that is available for plant uptake. In practice, at the scale at which we work, we generally know little about available effective irrigation water. At best we can obtain data on ratios of irrigated to non-irrigated cropland areas. In practice we are therefore forced to simply assume that throughout the growing period the amount of effective irrigation is sufficient to meet the irrigation requirement in all irrigated lands and that effective irrigation is zero in the case of non-irrigated lands. This implies that ET_b is assumed to equal IR for the irrigated areas and to be zero for the non-irrigated lands. In reality there are lands that are irrigated but not sufficiently to meet irrigation requirements at times, but this can only be dealt with if more detailed irrigation data are available.

Total evapotranspiration from the crop field is the sum of the two components calculated above. All above-mentioned water flows are expressed in mm/day, but in calculations we actually apply a time step of 5 days, to account for the possibility of soil moisture storage. Temporary storage of rain or irrigation water in the soil makes it possible that surplus water on one day can be used by the plants in the following few days, so that a day-by-day comparison of crop water requirement and effective rainfall or irrigation water would be too conservative.

The green and blue components in crop water use (CWU, m^3/ha) are calculated by accumulation of daily evapotranspiration over the complete growing period:

$$CWU_g = 10 \times \sum_{d=1}^{lgp} ET_g$$

$$CWU_b = 10 \times \sum_{d=1}^{lgp} ET_b$$

The factor 10 is to convert mm into m^3/ha. The summation is done over the period from the day of planting (day 1) to the day of harvest (lgp stands for length of growing period in days). Since different crop varieties can have substantial differences in the length of the growing period, this factor can significantly influence the calculated crop water use. For permanent crops we account for the evapotranspiration throughout the year. The "green" crop water use represents the total rainwater evaporation from the field during the growing period; the "blue" crop water use represents the total irrigation water evaporation from the field. Total crop water use – the sum of the above two components – is equal to the crop water requirements summed over the growing period if rainwater is sufficient throughout the growing period or if shortages are supplemented through irrigation.

The green component in the virtual-water content of a primary crop (v_g, m^3/ton) is calculated as the green component in crop water use (CWU_g, m^3/ha) divided by the crop yield Y (ton/ha). The blue component (v_b, m^3/ton) is calculated in a similar way, but should also include a component that refers to evaporation losses within the irrigation water storage and transport system:

$$v_g = \frac{CWU_g}{Y}$$
$$v_b = \frac{CWU_b}{Y} + \frac{E_{irr}}{Prod}$$

The symbol E_{irr} refers to the evaporation of irrigation water from artificial storage reservoirs and transport canals (m^3/yr), and $Prod$ to the total crop production (ton/yr). Evaporation of irrigation water even before it reaches the field can be considerable, but data are generally difficult to obtain. In this book we have not included this component in our studies, but we recommend including it in further studies.

The gray component in the virtual-water content of a primary crop (v_{gray}, m^3/ton) is calculated as the load of pollutants that enters the water system (L, kg/yr) divided by the maximum acceptable concentration for the pollutant considered (c_{max}, kg/m^3) and the crop production ($Prod$, ton/yr):

$$v_{gray} = \frac{L/c_{max}}{Prod}$$

In crop production, the pollutants will consist of for instance fertilizers (nitrogen, phosphorus) and pesticides. In the case of fertilizers one has to consider only the "waste flow" to the environment, which is only a fraction of the total application of fertilizers to the field. For calculating the gray component in the virtual-water content of the crop one needs to account for only the most critical pollutant, that is the pollutant where the above calculation yields the highest water volume. We did not include the gray component in the virtual-water content of crop products in our earlier studies.

In Chapters 2–8 one will therefore not be able to find this component. Only in Chapter 9, the study on cotton, do we add this component.

The total virtual-water content of a primary crop (v, m^3/ton) is the sum of the green, blue, and gray components:

$$v = v_g + v_b + v_{gray}$$

If a primary crop is processed into a crop product (e.g. sugar cane processed into raw sugar), there is often a loss in weight, because only part of the primary product is used. The rest is waste or at least relatively valueless. Let us label the primary crop as the root product r and the crop product as the processed product p. We then define the product fraction ($f_p[p]$, ton/ton) as the quantity of the processed product ($w[p]$, ton) obtained per quantity of root product ($w[r]$, ton):

$$f_p[p] = \frac{w[p]}{w[r]}$$

One can calculate the virtual-water content of the processed product by dividing the virtual-water content of the root product by the product fraction. If a primary crop is processed into two or more different products (e.g. soybean processed into soybean flour and soybean oil), one needs to distribute the virtual-water content of the primary crop across its separate products. This can be done proportionally to the value of the crop products. It could also be done proportionally to the weight of the products, but this would be less meaningful. We define the value fraction ($f_v[p]$, US$/US$) for a processed product p as the ratio of the market value of the product to the aggregated market value of all the products obtained from the root product:

$$f_v[p] = \frac{P[p] \times w[p]}{\sum\limits_{i=1}^{n}(p[i] \times w[i])}$$

Here, $P[p]$ refers to the market value (price) of product p (US$/ton). The denominator is summed over the n processed products that originate from the root product. If during processing there is some water use involved, the process water use is added to the virtual-water content of the root product before the total is distributed over the various processed products. In summary, the virtual-water content of a processed product is calculated as:

$$v[p] = (v[r] + PWU[r]) \times \frac{f_v[p]}{f_p[p]}$$

in which $v[p]$ is the virtual-water content (m³/ton) of product p, $v[r]$ the virtual-water content of the root product r, and $PWU[r]$ the process water use (m³/ton) when processing the root product into its derived products. In a similar way one can calculate the virtual-water content for products that result from a second or third processing step. The first step is always to obtain the virtual-water content of the root (input) product and the water used to process it. The total of these two elements is then distributed over the various processed (output) products, based on their product fraction and value fraction.

The virtual-water content of live animals can be calculated based on the virtual-water content of their feed and the volumes of drinking and service water consumed during their lifetime. Obviously, one will have to know the age of the animal when slaughtered and the diet of the animal during its various stages of life. The virtual-water content of livestock products (meat, eggs, milk, cheese, leather) can be based on product fractions and value fractions, following the above-described methodology. The procedure is most simple when an animal generates one type of product only, for example laying hens producing eggs: the total volume of water directly or indirectly consumed by the hen is attributed to the total number of eggs produced. In the case of more than one type of product from one animal, for example meat and leather from cattle, the procedure with product and value fractions can be used.

In the case of industrial products, the virtual-water content is expressed in this book in terms of m³/US$ instead of m³/ton. A practical problem with industrial products is that there is great heterogeneity of products and production methods. If one wishes to assess the virtual-water content of a specific product produced according to a particular method, one can relatively easily do that by studying the water use in the whole production chain and dividing the total water evaporation in the chain by the production value (to obtain the blue component of the virtual-water content), and dividing the load of pollutants to the water system by the maximum acceptable concentration and the total production value (to obtain the gray component). If the aim, however, is to rapidly assess the virtual-water content of all sorts of industrial products, one will have to consider broad product categories and the water use within whole sectors instead of within specific production chains. The blue water component in the virtual-water content of products from a particular industrial sector can be assessed by taking the ratio of the total water evaporation in this sector (m³/yr) to the generated added value of the sector (US$/yr). The gray component can be assessed as above, i.e. on the basis of the total load of pollutants from the sector, maximum acceptable concentrations, and the total production value. A practical problem is that data on evaporation and loads of pollutants are often not available. What we do often have, however, are data on total water withdrawals. In such cases a first crude estimate of the total virtual-water content of products from a sector can be obtained by dividing the total water withdrawal of the sector by the total production value. The rationale for this is that the water withdrawn either evaporates (the blue component in the virtual-water content) or returns in polluted form to the water system (the gray component).

The case of domestic water supply is similar to the case of industrial products. In fact, the water supply sector can be regarded as one specific industrial sector. Also here one can crudely assume that the total water volume withdrawn either evaporates or returns in polluted form to the water system. Treatment of wastewater flows before disposal into the natural water system can reduce the gray component in the virtual-water content of the product.

Virtual-Water Flows

International virtual-water flows can be calculated by multiplying commodity trade flows by their associated virtual-water content:

$$V[n_e, n_i, c] = T[n_e, n_i, c] \times v[n_e, c]$$

in which V denotes the virtual-water flow (m^3/yr) from exporting country n_e to importing country n_i as a result of trade in commodity c; T the commodity trade (ton/yr) from the exporting to the importing country; and v the virtual-water content (m^3/ton) of the commodity, which is defined as the volume of water used to produce the commodity in the exporting country. If the exporting country does not produce commodity c itself, but only imports it for further export, one should take the virtual-water content of the product as in the country of origin.

Water Savings Related to Trade

The national water saving S_n (m^3/yr) of a country n as a result of trade in product p has been defined as:

$$S_n[n, p] = (T_i[n, p] - T_e[n, p]) \times v[n, p]$$

where v is the virtual-water content (m^3/ton) of product p in country n, T_i the volume of product p imported (ton/yr), and T_e the volume of the product exported (ton/yr). Obviously, S_n can have a negative sign, which means a net water loss instead of a saving.

The global water saving S_g (m³/yr) through the trade in product p from an exporting country n_e to an importing country n_i is:

$$S_g[n_e, n_i, p] = T[n_e, n_i, p] \times (v[n_i, p] - v[n_e, p])$$

where T is the volume of trade (ton/yr) between the two countries. The global saving is thus obtained as the difference between the water productivities of the trading partners. When in a particular case the importing country is not able to produce the product domestically, we have taken the difference between the global average virtual-water content of the product and the virtual-water content in the exporting country.

The total global water saving can be obtained by adding together the global savings of all international trade flows. By definition, the total global water saving is equal to the sum of the national savings of all countries.

Water Footprints

The water footprint of a nation can be assessed through a bottom-up or a top-down approach. In the bottom-up approach, the water footprint of a nation is calculated by multiplying all goods and services consumed by the inhabitants of a country by the respective water needs for those goods and services:

$$WF_n = \sum_p^i C[p] \times v[p]$$

where C is the consumption in the country of product p (ton/yr) and v the virtual-water content of this product (m³/ton). The virtual-water content of a product should be taken as in the area of actual production. This means that consuming rice from Thailand will have a different contribution to the water footprint than consuming rice from Vietnam. Nobody has yet estimated a national water footprint by applying the bottom-up approach, but there is no reason to assume that it is not possible. In a sense, it is quite straightforward, if data demanding. We have however applied the bottom-up approach in assessing the water footprint of an individual. A simple calculator for assessing your individual water footprint, based on the bottom-up approach, is available at www.waterfootprint.org.

For assessing the water footprint of a nation it is easier to use the top-down approach, which takes total water use in a country as a starting point and then subtracts the part of the water used for making export products and adds the incoming virtual-water flow:

$$WF_n = WU_a + WU_i + WU_d + V_i - V_e$$

WU_a is the agricultural water use within the country, which is defined as the product of national crop production (ton/yr) and the average virtual-water content of the crop (m³/ton), totaled over the various crops. It is useful to observe here that the term "agricultural water use" is different from the term "agricultural water withdrawal," because the former includes three components in crop water use (blue, green, gray), while the latter includes only the blue component. A further difference is that the term "agricultural water use" excludes the part of the total water withdrawal that returns in unpolluted form to the natural water system. The symbol WU_i refers to the water use in the industrial sector, which is the production value (US$/yr) multiplied by the average virtual-water content of the industrial products (m³/US$), totaled over the various industrial sectors. WU_d refers to the water use in the domestic water supply sector. V_i is the gross virtual-water import and V_e is the gross virtual-water export. All flows are expressed in m³/yr.

A nation's water footprint has two components: the internal and the external water footprint. The internal water footprint (WF_i) is defined as the use of domestic water resources to produce goods and services consumed by inhabitants of the country. It is the sum of the total water volume used from the domestic water resources in the national economy minus the volume of virtual-water export to other countries insofar as related to export of domestically produced products:

$$WF_i = WU_a + WU_i + WU_d - V_{e,d}$$

$V_{e,d}$ is the part of the gross virtual-water export that concerns export of domestically produced products.

The external water footprint of a country (WF_e) is defined as the annual volume of water resources used in other countries to produce goods and services consumed by the inhabitants of the country concerned. It is equal to the virtual-water import into the country (V_i) minus the volume of virtual water exported to other countries as a result of re-export of imported products ($V_{e,r}$):

$$WF_e = V_i - V_{e,r}$$

Both internal and external water footprints include three components: the blue component (evaporated ground- or surface water), the green component (evaporated rainwater), and the gray component (polluted ground- or surface water).

Water Scarcity

We define water scarcity (WS) of a nation as the ratio of the nation's water footprint (WF_n) to the nation's water availability (WA):

$$WS = \frac{WF_n}{WA} \times 100$$

The national water scarcity can be more than 100% if there is more water needed for producing the goods and services consumed by the people of a nation than is available in the country. As a measure of water availability we take in this book (see Chapter 10 and Appendix V) the "total renewable water resources (actual)" as defined by the Food and Agriculture Organization in their AQUASTAT database (FAO, 2006a). One could argue however that it would be better to take the total precipitation in a country as a measure of water availability, because FAO's definition of renewable water resources measures blue water and thus excludes green water.

Water Dependency Versus Water Self-Sufficiency

We define the virtual-water import dependency (WD, %) of a nation as the ratio of the external water footprint (WF_e, m³/yr) to the total water footprint (WF_n, m³/yr) of a country:

$$WD = \frac{WF_e}{WF_n} \times 100$$

National water self-sufficiency (WSS, %) is defined as the internal water footprint (WF_i, m³/yr) divided by the total water footprint:

$$WSS = \frac{WF_i}{WF_n} \times 100$$

Self-sufficiency is 100% when all the water needed is available and indeed taken from within a nation's own territory. Water self-sufficiency approaches zero if the demands of goods and services in a country are largely met with gross virtual-water imports, i.e. the country has a relatively large external water footprint in comparison with its internal water footprint.

Appendix II

Virtual-Water Flows per Country Related to International Trade in Crop, Livestock, and Industrial Products. Period: 1997–2001

Country	Gross virtual-water flows (10^6 m³/yr)								Net virtual-water import (10^6 m³/yr)			
	Related to trade in crop products		Related to trade in livestock products		Related to trade in industrial products		Total		Related to trade in crop products	Related to trade in livestock products	Related to trade in industrial products	Total
	Export	Import	Export	Import	Export	Import	Export	Import				
Afghanistan	191	446	1	4	0	32	191	483	255	4	32	292
Albania	35	829	49	247	40	52	124	1,128	794	198	12	1,004
Algeria	441	10,980	43	1,099	394	442	879	12,521	10,539	1,056	48	11,642
Andorra	2	192	2	200	0	0	3	392	190	198	0	388
Angola	800	1,006	0	447	55	195	855	1,648	206	447	140	793
Anguilla	1	2	0	0	0	0	1	2	1	0	0	1
Antigua	28	10	2	9	0	21	30	40	-18	7	21	11
Argentina	45,952	3,100	4,178	811	499	1,732	50,629	5,643	-42,853	-3,367	1,233	-44,987
Armenia	24	492	3	138	40	45	67	675	469	134	5	608
Aruba	338	123	1	29	0	0	338	151	-215	28	0	-187
Australia	46,120	3,864	26,377	745	501	4,399	72,998	9,007	-42,256	-25,633	3,898	-63,991
Austria	1,996	5,578	2,852	2,040	1,248	4,504	6,095	12,122	3,582	-811	3,256	6,027
Azerbaijan	1,037	1,434	96	137	2,668	77	3,800	1,648	397	41	-2,591	-2,153
Bahamas	39	48	2	139	0	98	41	286	9	138	98	245
Bahrain	32	294	10	127	21	246	63	666	262	117	225	604
Bangladesh	771	3,670	652	86	162	415	1,585	4,171	2,899	-566	254	2,586
Barbados	107	118	8	94	7	60	122	273	11	85	54	151
Belarus	125	3,360	119	376	1,493	488	1,738	4,224	3,235	257	-1,005	2,486
Belgium-Luxembourg	14,688	31,658	9,825	4,494	17,703	10,955	42,217	47,106	16,970	-5,332	-6,749	4,889

Belize	456	23	5	26	17	21	478	70	−433	20	4	−408
Benin	1,937	386	8	28	4	34	1,948	448	−1,550	20	30	−1,500
Bermuda	33	105	4	28	0	38	37	170	72	24	38	133
Bhutan	6	84	0	6	1	9	8	100	78	6	8	92
Bolivia	1,858	711	395	53	12	111	2,265	875	−1,147	−342	99	−1,390
Bosnia-Herzegovina	38	891	134	1,097	0	205	172	2,192	853	962	205	2,020
Botswana	8	384	154	92	56	110	217	586	376	−62	54	369
Brazil	53,713	17,467	11,911	1,907	2,211	3,694	67,835	23,068	−36,246	−10,003	1,483	−44,767
Brunei	12	316	2	32	0	88	14	436	304	30	88	422
Bulgaria	1,778	1,203	423	500	9,641	372	11,843	2,075	−575	77	−9,269	−9,767
Burkina Faso	1,544	312	10	20	0	36	1,554	368	−1,232	9	36	−1,186
Burundi	329	130	0	2	0	8	330	140	−199	1	8	−190
Cambodia	25	418	24	34	0	89	49	541	394	10	89	492
Cameroon	8,300	1,121	4	37	38	85	8,342	1,243	−7,179	33	48	−7,099
Canada	48,321	16,190	17,424	4,952	29,573	14,289	95,318	35,430	−32,132	−12,472	−15,284	−59,888
Cape Verde Islands	2	51	0	5	0	10	2	65	49	5	10	63
Cayman Islands	8	216	0	27	0	39	9	281	207	27	39	273
Central African Republic	650	47		1	0	9	650	57	−603	1	9	−593
Chad	1,960	64	6	2	0	27	1,966	94	−1,895	−4	27	−1,872
Chile	1,122	3,415	265	1,186	1,051	1,182	2,438	5,783	2,293	921	131	3,345
China	17,429	36,260	5,640	15,247	49,909	11,632	72,978	63,139	18,831	9,608	−38,277	−9,839
Cocos Islands	26	2	1	0	0	0	27	2	−24	−1	0	−25
Colombia	10,783	6,169	487	310	127	779	11,397	7,258	−4,614	−177	652	−4,139
Comoros	72	152	0	67	0	4	72	223	81	66	4	151

(Continued)

Country	Gross virtual-water flows (10^6 m³/yr)								Net virtual-water import (10^6 m³/yr)			
	Related to trade in crop products		Related to trade in livestock products		Related to trade in industrial products		Total		Related to trade in crop products	Related to trade in livestock products	Related to trade in industrial products	Total
	Export	Import	Export	Import	Export	Import	Export	Import				
Congo	259	396	0	107	0	59	259	561	136	107	59	302
Congo, Democratic Republic	796	328	0	73	0	76	797	477	−469	73	76	−320
Costa Rica	2,979	1,262	365	102	308	388	3,652	1,753	−1,717	−263	80	−1,899
Côte d'Ivoire	35,029	2,048	13	119	47	161	35,089	2,328	−32,981	106	114	−32,761
Croatia	464	1,254	308	1,284	0	526	772	3,064	790	976	526	2,292
Cuba	8,628	2,223	15	317	28	247	8,671	2,787	−6,405	302	218	−5,884
Cyprus	245	1,012	67	62	1	205	314	1,279	767	−5	204	965
Czech Republic	2,039	3,845	867	683	1,722	1,957	4,628	6,486	1,806	−184	236	1,858
Denmark	2,696	6,460	9,462	1,567	262	2,693	12,419	10,719	3,764	−7,895	2,431	−1,700
Djibouti	10	307	5	17	0	17	15	341	296	13	17	326
Dominica	50	28	5	9	0	7	55	44	−22	4	7	−11
Dominican Republic	3,309	0	67	0	35	495	3,410	495	−3,309	−67	460	−2,916
Ecuador	7,385	1,426	67	48	341	278	7,792	1,752	−5,959	−18	−63	−6,040
Egypt	1,755	11,445	221	1,466	729	711	2,705	13,622	9,690	1,245	−18	10,917
El Salvador	2,718	1,346	72	405	98	257	2,888	2,008	−1,372	333	159	−880
Equatorial Guinea	476	56	0	13	0	31	476	100	−420	13	31	−376
Eritrea	14	238	18	7	0	27	31	272	225	−11	27	241
Estonia	399	2,962	162	394	112	277	673	3,634	2,563	232	165	2,961

Ethiopia	2,143	346	90	2	5	89	2,238	437	-1,797	-88	83	-1,801
Faeroes	1	31	0	21	0	24	1	77	30	21	24	75
Falkland Islands	14	3	0	1	0	0	14	4	-11	1	0	-10
Fiji	564	0	9	0	3	49	577	49	-564	-9	47	-527
Finland	1,015	2,737	569	355	1,827	2,087	3,412	5,179	1,722	-214	260	1,767
France	43,410	40,577	13,222	11,829	21,873	19,761	78,505	72,166	-2,833	-1,393	-2,112	-6,338
French Polynesia	21	80	0	108	0	53	21	241	58	108	53	220
Gabon	70	328	0	112	14	57	84	497	258	112	43	413
Gambia	142	479	1	21	0	8	142	509	337	21	8	366
Georgia	347	242	217	23	204	43	768	308	-105	-193	-162	-460
Germany	27,630	59,751	17,432	16,062	25,416	29,757	70,478	1,05,570	32,121	-1,370	4,341	35,092
Ghana	19,501	1,482	2	69	14	169	19,516	1,720	-18,019	67	155	-17,797
Gibraltar	1	24	2	20	0	0	3	44	23	18	0	42
Greece	4,634	4,548	336	4,255	243	1,785	5,212	10,588	-86	3,919	1,543	5,375
Greenland	3	13	1	25	0	0	4	38	9	24	0	34
Grenada	176	32	0	18	0	11	176	62	-144	18	11	-114
Guatemala	5,684	1,416	166	350	60	277	5,911	2,043	-4,268	183	217	-3,868
Guinea	886	859	2	20	10	30	898	909	-27	18	20	11
Guinea-Bissau	33	52	0	3	0	5	33	59	19	3	5	26
Guyana	1,033	94	1	24	0	41	1,034	159	-940	23	41	-876
Haiti	253	0	5	0	1	63	258	63	-253	-5	62	-195
Honduras	3,043	517	77	135	29	149	3,149	801	-2,526	58	119	-2,348
Hong Kong	304	8,773	380	18,790	0	0	684	27,563	8,469	18,410	0	26,878
Hungary	3,495	2,795	8,586	628	0	0	12,081	3,423	-700	-7,958	0	-8,658
Iceland	9	165	62	2	27	147	98	314	156	-60	120	216
India	32,411	13,941	3,406	343	6,748	2,945	42,565	17,228	-18,470	-3,063	-3,803	-25,337

(Continued)

Country	Gross virtual-water flows (10^6 m³/yr)								Net virtual-water import (10^6 m³/yr)			
	Related to trade in crop products		Related to trade in livestock products		Related to trade in industrial products		Total		Related to trade in crop products	Related to trade in livestock products	Related to trade in industrial products	Total
	Export	Import	Export	Import	Export	Import	Export	Import				
Indonesia	24,750	26,917	371	1,666	310	1,822	25,430	30,405	2,167	1,296	1,512	4,975
Iran	3,587	18,150	314	474	911	817	4,812	19,442	14,563	160	-94	14,630
Iraq	703	2,901	5	280	0	581	707	3,762	2,198	275	581	3,055
Ireland	623	3,385	6,239	839	1,116	2,931	7,978	7,155	2,762	-5,400	1,815	-823
Israel	575	4,111	140	687	71	2,156	786	6,954	3,537	547	2,084	6,168
Italy	12,920	47,164	14,912	28,295	10,402	13,498	38,234	88,957	34,244	13,383	3,096	50,723
Jamaica	489	621	11	199	28	180	529	999	132	187	151	470
Japan	954	59,015	955	20,328	4,605	18,883	6,513	98,227	58,061	19,374	14,279	91,714
Jordan	97	4,103	165	462	25	228	287	4,794	4,006	297	203	4,506
Kazakhstan	7,363	317	648	70	4,819	295	12,830	682	-7,046	-578	-4,524	-12,148
Kenya	4,638	2,361	161	13	28	182	4,828	2,555	-2,277	-149	154	-2,272
Kiribati	1	13	0	7	0	2	1	21	12	7	2	20
Korea Republic	997	24,801	3,930	6,097	2,219	8,344	7,146	39,242	23,804	2,166	6,126	32,096
Korea, Democratic People's Republic	31	1,915	19	67	0	99	50	2,081	1,884	48	99	2,031
Kuwait	30	1,058	23	441	8	460	61	1,960	1,028	419	452	1,899
Kyrgyzstan	296	0	128	0	135	37	558	37	-296	-128	-98	-521
Laos	246	161	22	10	0	40	268	211	-85	-12	40	-57
Latvia	387	661	192	148	61	185	640	995	275	-44	124	355

Lebanon	212	2,744	75	1,379	4	380	291	4,503	2,532	1,304	376	4,212
Liberia	195	167	0	25	0	40	195	233	−28	25	40	38
Libya	102	3,372	51	404	80	329	233	4,105	3,270	353	249	3,872
Lithuania	339	1,229	591	291	37	341	967	1,861	890	−300	304	894
Macau	3	160	5	121	0	0	8	281	156	117	0	273
Macedonia	154	555	37	337	0	107	191	1,000	401	300	107	809
Madagascar	3,249	291	28	6	0	48	3,277	345	−2,958	−22	48	−2,932
Malawi	778	132	2	9	4	36	784	177	−646	8	32	−607
Malaysia	23,881	16,720	1,012	2,439	3,286	4,663	28,179	23,822	−7,161	1,427	1,377	−4,357
Maldives	2	190	0	234	0	20	2	443	187	234	20	441
Mali	3,368	92	9	6	0	48	3,378	145	−3,277	−4	47	−3,233
Malta	22	317	2	177	1	175	26	669	295	175	174	644
Marshall Islands	16	13	0	7	0	4	16	24	−3	7	4	8
Mauritania	2	945	2	21	0	25	4	992	944	20	25	988
Mauritius	647	1,033	33	258	48	121	728	1,412	386	225	73	684
Mexico	11,784	26,956	5,757	13,418	3,790	9,710	21,331	50,084	15,173	7,661	5,920	28,754
Micronesia	4	26	0	19	0	4	4	50	22	19	4	45
Moldavia	1,390	133	304	43	887	54	2,582	230	−1,257	−261	−833	−2,351
Mongolia	8	73	1,744	2	0	0	1,752	74	64	−1,742	0	−1,678
Montserrat	9	1	0	0	0	1	9	2	−8	0	1	−6
Morocco	1,327	6,127	225	164	123	599	1,676	6,891	4,800	−61	476	5,216
Mozambique	1,112	0	6	0	3	57	1,121	57	−1,112	−6	54	−1,064
Myanmar	1,447	885	100	72	0	171	1,547	1,128	−562	−28	171	−419
N. Mariana	7	75	14	35	0	55	21	166	69	21	55	145
Namibia	3	48	99	3	9	97	111	147	45	−96	88	37
Nauru	1	2	0	6	0	1	2	9	0	6	1	7

(Continued)

Country	Gross virtual-water flows (10⁶ m³/yr)								Net virtual-water import (10⁶ m³/yr)			
	Related to trade in crop products		Related to trade in livestock products		Related to trade in industrial products		Total		Related to trade in crop products	Related to trade in livestock products	Related to trade in industrial products	Total
	Export	Import	Export	Import	Export	Import	Export	Import				
Nepal	140	587	51	16	11	88	203	691	447	−36	77	488
Netherland Antilles	152	113	7	117	0	0	158	230	−38	110	0	72
Netherlands	34,529	48,607	15,146	7,852	7,885	12,293	57,561	68,753	14,078	−7,294	4,408	11,192
New Zealand	327	1,600	8,984	300	70	871	9,381	2,770	1,273	−8,684	801	−6,611
Nicaragua	1,721	655	537	100	5	97	2,262	852	−1,066	−436	92	−1,410
Niger	120	592	7	12	6	17	132	622	473	5	11	490
Nigeria	8,673	5,541	64	253	531	542	9,268	6,335	−3,132	189	11	−2,933
Norway	257	2,619	276	168	1,081	2,219	1,615	5,006	2,361	−108	1,138	3,391
Oman	200	2,250	53	590	20	290	272	3,130	2,050	537	271	2,858
Pakistan	7,381	8,879	612	98	1,526	579	9,518	9,555	1,498	−514	−947	37
Panama	658	566	207	101	5	199	869	866	−92	−106	195	−3
Papua New Guinea	6,392	8,054	6	159	33	73	6,432	8,286	1,662	153	40	1,854
Paraguay	5,311	231	911	58	3	143	6,225	432	−5,079	−853	140	−5,793
Peru	2,403	4,410	24	404	404	458	2,831	5,272	2,007	380	54	2,441
Philippines	8,454	10,115	54	2,524	3,184	2,208	11,691	14,846	1,661	2,471	−976	3,155
Poland	1,927	10,425	2,190	1,321	5,267	2,969	9,383	14,715	8,499	−869	−2,298	5,332
Portugal	2,633	9,631	780	2,856	1,320	2,287	4,733	14,774	6,998	2,076	967	10,041
Qatar	14	195	9	63	7	197	29	455	181	54	190	425
Romania	2,187	2,538	668	1,799	5,199	789	8,054	5,126	351	1,130	−4,410	−2,929

Russia	8,297	30,925	2,503	12,243	36,932	2,899	47,732	46,067	22,627	9,740	−34,032	−1,665
Rwanda	219	255	4	7	0	13	224	275	36	2	13	51
Samoa	19	15	0	27	0	8	19	50	−3	27	8	31
Sao Tome and Principe	203	24	1	209	0	1	205	234	−179	207	1	29
Saudi Arabia	362	10,598	556	2,006	136	1,703	1,054	14,308	10,236	1,450	1,568	13,254
Senegal	3,201	3,418	22	65	19	74	3,242	3,557	217	43	55	315
Seychelles	37	104	0	16	0	25	37	144	67	15	25	107
Sierra Leone	95	139	0	10	0	8	95	157	44	9	8	62
Singapore	4,852	7,238	364	1,825	0	7,934	5,216	16,997	2,386	1,461	7,934	11,781
Slovakia	728	1,444	304	610	0	817	1,032	2,871	716	306	817	1,840
Slovenia	163	1,271	393	666	0	623	556	2,560	1,108	273	623	2,004
Solomon Islands	147	241	0	4	0	8	147	254	94	4	8	107
Somalia	43	730	129	2	0	13	172	744	686	−127	13	572
South Africa	6,326	7,752	1,312	1,019	912	1,924	8,550	10,695	1,426	−293	1,011	2,145
Spain	18,252	30,483	8,541	5,972	3,753	8,520	30,545	44,975	12,231	−2,569	4,767	14,430
Sri Lanka	2,381	1,373	46	83	213	366	2,640	1,822	−1,008	37	153	−818
St Helena	0	44	0	1	0	0	0	45	44	1	0	45
St Kitts	39	8	0	11	0	9	39	28	−31	11	9	−11
St Lucia	111	37	0	22	0	18	111	77	−74	22	18	−34
St Vincent	108	51	0	12	0	9	108	73	−56	12	9	−34
St. Piere Maq.	5	1	0	1	0	5	5	7	−4	0	5	1
Sudan	7,251	520	273	10	56	89	7,580	619	−6,730	−263	33	−6,960
Surinam	178	21	1	5	39	34	219	60	−157	4	−5	−158
Swaziland	0	134	0	41	21	62	21	237	134	41	41	216

(Continued)

Country	Gross virtual-water flows (10⁶ m³/yr)								Net virtual-water import (10⁶ m³/yr)			
	Related to trade in crop products		Related to trade in livestock products		Related to trade in industrial products		Total		Related to trade in crop products	Related to trade in livestock products	Related to trade in industrial products	Total
	Export	Import	Export	Import	Export	Import	Export	Import				
Sweden	2,034	4,737	808	1,203	1,639	4,316	4,481	10,256	2,703	395	2,678	5,776
Switzerland-Liechtenstein	1,163	6,172	401	752	1,555	5,208	3,119	12,132	5,008	351	3,654	9,013
Syria	4,025	3,131	512	143	126	213	4,664	3,488	-894	-368	87	-1,176
Taiwan	329	11,708	3,559	3,535	0	0	3,888	15,243	11,380	-24	0	11,355
Tajikistan	1,014	0	36	0	0	0	1,049	0	-1,014	-36	0	-1,049
Tanzania	3,173	970	52	11	2	85	3,227	1,066	-2,203	-41	83	-2,161
Thailand	38,429	9,761	2,856	1,761	1,655	3,596	42,940	15,117	-28,668	-1,096	1,941	-27,823
Togo	1,920	400	2	8	7	34	1,929	443	-1,519	6	27	-1,486
Tokelau Islands	2	1	0	1	0	0	2	2	-1	1	0	0
Trinidad and Tobago	350	493	15	169	81	193	446	854	143	153	112	409
Tunisia	11,013	3,502	78	211	72	524	11,162	4,237	-7,510	133	452	-6,925
Turkey	11,069	14,069	337	1,206	1,902	2,941	13,308	18,216	3,000	869	1,040	4,908
Turkmenistan	1,071	165	27	43	72	92	1,170	301	-906	16	21	-869
Tuvalu	0	1	0	1	0	0	0	2	1	1	0	2
Uganda	4,432	1,201	77	3	1	88	4,511	1,293	-3,231	-74	87	-3,218
UK	8,773	33,742	3,786	10,163	5,113	20,321	17,672	64,226	24,968	6,378	15,208	46,554

Ukraine	8,154	2,205	2,447	1,025	10,414	965	21,016	4,195	−5,949	−1,423	−9,449	−16,821
United Arab Emirates	4,603	0	475	0	167	1,718	5,245	1,718	−4,603	−475	1,551	−3,527
Uruguay	2,009	555	3,356	159	6	209	5,371	923	−1,454	−3,197	203	−4,448
USA	1,34,623	73,129	35,484	32,919	59,195	69,763	2,29,303	1,75,811	−61,495	−2,564	10,568	−53,491
Uzbekistan	6,533	1,195	55	209	0	229	6,588	1,634	−5,338	154	229	−4,954
Vanatu	138	0	22	0	0	5	160	5	−138	−22	5	−155
Venezuela	1,394	4,931	389	435	396	962	2,179	6,328	3,537	46	566	4,149
Vietnam	11,124	2,278	165	291	0	848	11,289	3,417	−8,846	126	848	−7,872
Virgin Islands	21	124	0	3	0	14	21	141	103	3	14	120
Yemen	243	3,638	37	168	17	151	297	3,957	3,395	131	133	3,659
Yugoslavia	518	942	144	0	0	255	662	1,197	424	−144	255	535
Zambia	508	237	14	0	101	63	623	299	−271	−14	−38	−323
Zimbabwe	3,032	0	319	0	43	146	3,394	146	−3,032	−319	103	−3,247
Others	110	3,356	1	1,815	0	6,993	111	12,164	3,247	1,813	6,993	12,053
Total	9,86,259	9,86,259	2,76,222	2,76,222	3,61,838	3,61,838	16,24,319	16,24,319	0	0	0	0

Appendix III

National Water Savings and Losses Due to Trade in Agricultural Products. Period: 1997–2001

Country	Gross saving (10^9 m^3/yr)	Gross loss (10^9 m^3/yr)	Net saving (10^9 m^3/yr)
Afghanistan	0.8	0.2	0.6
Albania	1.7	0.1	1.6
Algeria	45.9	0.5	45.4
Andorra	0.4	0.0	0.4
Angola	3.6	0.8	2.8
Anguilla	0.0	0.0	0.0
Antigua	0.0	0.0	0.0
Argentina	3.2	50.1	−46.9
Armenia	0.7	0.0	0.7
Aruba	0.0	0.3	−0.3
Australia	6.2	63.4	−57.2
Austria	7.4	4.8	2.6
Azerbaijan	1.4	1.1	0.2
Bahamas	0.3	0.0	0.3
Bahrain	0.4	0.0	0.4
Bangladesh	4.3	1.4	2.9
Barbados	0.3	0.1	0.2
Belarus	2.7	0.2	2.5
Belgium-Luxembourg	43.3	24.5	18.8
Belize	0.1	0.5	−0.4
Benin	0.7	1.9	−1.2
Bermuda	0.2	0.0	0.1
Bhutan	0.2	0.0	0.1
Bolivia	2.5	2.3	0.2
Bosnia-Herzegovina	2.8	0.2	2.6
Botswana	1.7	0.2	1.5
Brazil	29.6	65.6	−36.0
Brunei	0.5	0.0	0.4
Bulgaria	2.9	2.2	0.7
Burkina Faso	0.5	1.6	−1.1
Burundi	0.1	0.3	−0.2
Cambodia	0.9	0.0	0.8
Cameroon	1.8	8.3	−6.5
Canada	23.1	65.7	−42.7
Cape Verde Islands	0.2	0.0	0.2
Cayman Islands	0.3	0.0	0.3
Central African Republic	0.1	0.6	−0.5
Chad	0.2	2.0	−1.8

Country	Gross saving (10^9 m³/yr)	Gross loss (10^9 m³/yr)	Net saving (10^9 m³/yr)
Chile	3.9	1.4	2.6
China	50.6	23.1	27.5
(excluding Hong Kong)			
Cocos Islands	0.0	0.0	0.0
Colombia	9.3	11.2	−1.9
Comoros	0.4	0.1	0.3
Congo	0.9	0.3	0.6
Congo, Democratic Republic	0.9	0.8	0.2
Costa Rica	2.4	3.3	−0.9
Côte d'Ivoire	2.8	35.0	−32.2
Croatia	2.7	0.8	1.9
Cuba	4.2	8.6	−4.5
Cyprus	1.8	0.3	1.4
Czech Republic	5.2	2.9	2.3
Denmark	9.0	12.1	−3.2
Djibouti	0.4	0.0	0.4
Dominica	0.1	0.1	0.0
Dominican Republic	0.0	3.4	−3.4
Ecuador	4.0	7.5	−3.5
Egypt	15.2	2.0	13.2
El Salvador	2.6	2.8	−0.2
Equatorial Guinea	0.1	0.5	−0.4
Eritrea	1.3	0.0	1.2
Estonia	3.2	0.6	2.6
Ethiopia	0.8	2.2	−1.4
Faeroes	0.1	0.0	0.1
Falkland Islands	0.0	0.0	0.0
Fiji	0.0	0.6	−0.6
Finland	3.7	1.6	2.2
France	48.1	56.6	−8.5
Gabon	0.6	0.1	0.5
Gambia	1.0	0.1	0.8
Georgia	0.4	0.6	−0.2
Germany	79.0	45.1	33.9
Ghana	2.4	19.5	−17.1
Gibraltar	0.0	0.0	0.0
Greece	11.3	5.0	6.3

(*Continued*)

Country	Gross saving $(10^9 \ m^3/yr)$	Gross loss $(10^9 \ m^3/yr)$	Net saving $(10^9 \ m^3/yr)$
Greenland	0.0	0.0	0.0
Grenada	0.1	0.2	−0.1
Guatemala	2.8	5.9	−3.0
Guinea	2.3	0.9	1.4
Guinea-Bissau	0.1	0.0	0.1
Guyana	0.2	1.0	−0.8
Haiti	0.0	0.3	−0.3
Honduras	2.0	3.1	−1.1
Hong Kong	28.7	0.7	28.0
Hungary	6.9	12.1	−5.1
Iceland	0.2	0.1	0.1
India	22.9	35.8	−12.9
Indonesia	28.2	25.1	3.1
Iran	40.8	3.9	36.9
Iraq	16.0	0.7	15.3
Ireland	4.0	6.9	−2.9
Israel	10.5	0.7	9.8
Italy	87.2	27.8	59.3
Jamaica	1.6	0.5	1.1
Japan	96.0	1.9	94.1
Jordan	8.4	0.3	8.1
Kazakhstan	0.6	8.0	−7.4
Kenya	3.5	4.8	−1.3
Kiribati	0.0	0.0	0.0
Korea Republic	38.5	4.9	33.6
Korea, Democratic People's Republic	2.9	0.1	2.9
Kuwait	2.2	0.1	2.1
Kyrgyzstan	0.0	0.4	−0.4
Laos	0.2	0.3	−0.1
Latvia	1.3	0.6	0.7
Lebanon	5.5	0.3	5.2
Liberia	0.5	0.2	0.3
Libya	18.0	0.2	17.8
Lithuania	1.3	0.9	0.4
Macau	0.3	0.0	0.2
Macedonia	1.1	0.2	0.9
Madagascar	0.4	3.3	−2.8

Country	Gross saving $(10^9 \ m^3/yr)$	Gross loss $(10^9 \ m^3/yr)$	Net saving $(10^9 \ m^3/yr)$
Malawi	0.3	0.8	−0.5
Malaysia	19.9	24.9	−5.0
Maldives	0.2	0.0	0.2
Mali	0.3	3.4	−3.1
Malta	0.5	0.0	0.5
Marshall Islands	0.0	0.0	0.0
Mauritania	3.3	0.0	3.2
Mauritius	1.1	0.7	0.4
Mexico	83.1	17.5	65.5
Micronesia	0.1	0.0	0.1
Moldavia	0.2	1.7	−1.5
Mongolia	0.4	1.8	−1.4
Montserrat	0.0	0.0	0.0
Morocco	28.6	1.6	27.0
Mozambique	0.0	1.1	−1.1
Myanmar	1.6	1.5	0.0
N. Mariana	0.0	0.0	0.0
Namibia	0.1	0.1	0.0
Nauru	0.0	0.0	0.0
Nepal	0.7	0.2	0.5
Netherland Antilles	0.1	0.2	0.0
Netherlands	57.7	49.6	8.1
New Zealand	1.8	9.3	−7.6
Nicaragua	1.1	2.3	−1.2
Niger	1.3	0.1	1.2
Nigeria	14.1	8.7	5.3
Norway	3.3	0.5	2.7
Oman	2.3	0.3	2.0
Pakistan	18.6	8.0	10.7
Panama	1.5	0.9	0.6
Papua New Guinea	0.1	6.4	−6.2
Paraguay	0.5	6.2	−5.7
Peru	7.6	2.4	5.2
Philippines	16.3	8.5	7.8
Poland	11.2	4.1	7.1
Portugal	14.2	3.4	10.8
Qatar	0.2	0.0	0.2
Romania	4.5	2.9	1.6

(*Continued*)

Country	Gross saving (10⁹ m³/yr)	Gross loss (10⁹ m³/yr)	Net saving (10⁹ m³/yr)
Russia	51.7	10.8	40.9
Rwanda	0.5	0.2	0.3
Samoa	0.1	0.0	0.1
Sao Tome and Principe	0.1	0.2	−0.1
Saudi Arabia	21.5	0.7	20.9
Senegal	6.4	3.2	3.2
Seychelles	0.1	0.0	0.1
Sierra Leone	0.3	0.1	0.2
Singapore	10.1	5.2	4.9
Slovakia	2.1	1.0	1.0
Slovenia	1.7	0.6	1.1
Solomon Islands	0.2	0.1	0.1
Somalia	2.8	0.2	2.6
South Africa	11.7	7.6	4.1
Spain	50.0	26.8	23.2
Sri Lanka	1.8	2.4	−0.6
St Helena	0.0	0.0	0.0
St Kitts	0.0	0.0	0.0
St Lucia	0.1	0.1	0.0
St. Piere Maq.	0.0	0.0	0.0
Sudan	1.1	7.5	−6.4
Surinam	0.0	0.2	−0.1
Sweden	6.5	2.8	3.7
Switzerland-Liechtenstein	6.3	1.6	4.8
Syria	16.1	4.5	11.5
Taiwan	17.4	3.9	13.5
Tajikistan	0.0	1.0	−1.0
Tanzania	1.7	3.2	−1.5
Thailand	15.6	41.2	−25.5
Togo	0.6	1.9	−1.3
Tokelau Islands	0.0	0.0	0.0
Trinidad and Tobago	1.0	0.4	0.7
Tunisia	7.2	11.1	−3.9
Turkey	18.1	11.4	6.7
Turkmenistan	0.3	1.1	−0.8
Tuvalu	0.0	0.0	0.0
Uganda	2.6	4.5	−1.9
UK	45.4	12.6	32.9

Country	Gross saving (10^9 m^3/yr)	Gross loss (10^9 m^3/yr)	Net saving (10^9 m^3/yr)
Ukraine	2.7	10.6	−7.9
United Arab Emirates	0.0	5.1	−5.1
Uruguay	0.8	5.4	−4.5
USA	78.1	169.9	−91.8
Uzbekistan	1.5	6.6	−5.1
Vanatu	0.0	0.2	−0.2
Venezuela	16.5	1.8	14.7
Vietnam	2.9	11.3	−8.4
Yemen	8.6	0.3	8.4
Yugoslavia	1.1	0.7	0.5
Zambia	0.4	0.5	−0.1
Zimbabwe	0.0	3.4	−3.4
Others	1.1	0.9	0.2
Global total	1,605	1,253	352

Appendix IV

Water Footprints of Nations.
Period: 1997–2001

Country	Population (1,000s)	Use of domestic water resources (10⁹ m³/yr)						Use of foreign water resources (10⁹ m³/yr)			Water footprint		Water footprint by consumption category (m³/cap/yr)				
		Domestic water withdrawal	Crop evapo-transpiration		Industrial water withdrawal			For national consumption		For re-export of imported products	Total (10^9 m³/yr)	Per capita (m³/cap/yr)	Consumption of domestic water	Consumption of agricultural goods		Consumption of industrial goods	
			For national consumption	For export	For national consumption	For export		Agricultural goods	Industrial goods				Internal water footprint	Internal water footprint	External water footprint	Internal water footprint	External water footprint
Afghanistan	26,179	0.34	16.47	0.19	0.001	0.00		0.45	0.03	0.01	17.29	660	13	629	17	0	1
Albania	3,131	0.24	2.43	0.06	0.086	0.03		1.05	0.04	0.04	3.84	1,228	75	777	336	28	13
Algeria	30,169	1.23	22.77	0.32	0.494	0.25		11.91	0.29	0.31	36.69	1,216	41	755	395	16	10
Angola	12,953	0.07	11.37	0.71	0.041	0.01		1.37	0.15	0.13	13.00	1,004	6	878	106	3	12
Argentina	36,806	4.68	41.31	48.03	2.328	0.30		1.81	1.53	2.30	51.66	1,404	127	1,122	49	63	42
Armenia	3,131	0.87	1.19	0.02	0.095	0.03		0.62	0.03	0.03	2.81	898	279	379	198	30	11
Australia	19,072	6.51	14.03	68.67	1.229	0.12		0.78	4.02	4.21	26.56	1,393	341	736	41	64	211
Austria	8,103	0.76	2.98	1.89	1.070	0.29		4.66	3.55	3.91	13.02	1,607	94	368	575	132	438
Azerbaijan	8,015	0.80	3.88	0.85	1.824	2.62		1.29	0.03	0.33	7.83	977	100	485	161	228	4
Bahrain	647	0.11	0.04	0.00	0.009	0.08		0.38	0.23	0.06	0.77	1,184	165	64	593	14	348
Bangladesh	129,943	2.12	109.98	1.38	0.344	0.00		3.71	0.34	0.13	116.49	896	16	846	29	3	3
Barbados	267	0.05	0.10	0.05	0.017	0.00		0.14	0.06	0.07	0.36	1,355	169	374	540	64	208
Belarus	10,020	0.63	8.21	0.17	0.166	0.00		3.66	0.07	0.50	12.74	1,271	63	820	365	17	7
Belgium-Luxembourg	10,659	1.09	2.29	3.26	0.382	7.29		14.90	0.54	31.66	19.21	1,802	103	215	1,398	36	51
Belize	236	0.01	0.26	0.43	0.083	0.01		0.02	0.02	0.03	0.39	1,646	53	1,087	77	352	77
Benin	6,192	0.04	10.47	1.88	0.019	0.00		0.35	0.03	0.07	10.91	1,761	6	1,690	57	3	5
Bhutan	793	0.01	0.72	0.01	0.003	0.00		0.09	0.01	0.00	0.83	1,044	14	902	113	4	10
Bolivia	8,233	0.15	9.02	2.11	0.037	0.00		0.62	0.10	0.15	9.93	1,206	19	1,095	75	4	12
Botswana	1,658	0.04	0.50	0.09	0.015	0.01		0.40	0.06	0.12	1.03	623	27	304	244	9	39
Brazil	169,110	11.76	195.29	61.01	8.666	1.63		14.76	3.11	5.20	233.59	1,381	70	1,155	87	51	18
Bulgaria	8,126	0.37	9.50	1.92	0.048	9.27		1.42	0.00	0.66	11.33	1,395	45	1,169	174	6	0
Burkina Faso	11,138	0.08	16.61	1.53	0.001	0.00		0.30	0.04	0.03	17.03	1,529	7	1,491	27	0	3
Burundi	6,743	0.04	6.98	0.32	0.001	0.00		0.13	0.01	0.01	7.16	1,062	6	1,036	19	0	1
Cambodia	11,885	0.05	20.39	0.05	0.014	0.00		0.45	0.09	0.00	20.99	1,766	4	1,715	38	1	7
Cameroon	14,718	0.18	15.02	7.91	0.060	0.02		0.76	0.07	0.42	16.09	1,093	12	1,021	52	4	4
Canada	30,650	8.55	30.22	52.34	11.211	20.36		7.74	5.07	22.62	62.80	2,049	279	986	252	366	166
Cape Verde	429	0.00	0.36	0.00	0.001	0.00		0.05	0.01	0.00	0.43	995	8	835	128	1	23

Country																
Central African Republic	3,689	0.02	3.93	0.64	0.004	0.00	0.04	0.01	0.01	4.00	1,083	4	1,064	11	1	2
Chad	7,595	0.04	14.90	1.96	0.003	0.00	0.06	0.03	0.01	15.03	1,979	5	1,962	8	0	4
Chile	15,113	1.26	4.13	0.71	1.955	0.73	3.93	0.86	1.00	12.13	803	83	274	260	129	57
China	1,257,521	33.32	711.10	21.55	81.531	45.73	49.99	7.45	5.69	883.39	702	26	565	40	65	6
Colombia	41,919	5.31	23.08	9.40	0.358	0.04	4.60	0.70	1.96	34.05	812	127	551	110	9	17
Congo, Democratic Republic	50,265	0.20	36.16	0.79	0.058	0.00	0.39	0.08	0.01	36.89	734	4	719	8	1	2
Costa Rica	3,767	0.77	2.41	2.63	0.262	0.16	0.65	0.24	0.86	4.33	1,150	205	639	173	70	64
Côte d'Ivoire	15,792	0.19	26.71	33.83	0.077	0.02	0.96	0.13	1.24	28.06	1,777	12	1,692	61	5	8
Cuba	11,175	2.05	14.70	7.76	0.481	0.02	1.66	0.24	0.89	19.13	1,712	184	1,315	149	43	21
Cyprus	755	0.06	0.52	0.12	0.004	0.00	0.88	0.20	0.20	1.67	2,208	77	693	1,163	6	270
Czech Republic	10,269	1.08	9.59	2.09	0.763	0.75	3.72	0.99	1.78	16.15	1,572	106	934	362	74	96
Denmark	5,330	0.38	2.36	6.31	0.296	0.03	2.18	2.46	6.08	7.68	1,440	72	442	409	56	461
Dominican Republic	8,305	1.00	6.63	3.38	0.046	0.00	0.00	0.46	0.03	8.14	980	121	798	0	6	56
Ecuador	12,528	2.11	11.37	6.89	0.666	0.26	0.92	0.20	0.64	15.26	1,218	168	907	73	53	16
Egypt	63,376	4.16	45.78	1.55	6.423	0.66	12.49	0.64	0.49	69.50	1,097	66	722	197	101	10
El Salvador	6,216	0.28	3.69	2.15	0.134	0.04	1.11	0.20	0.70	5.41	870	45	593	178	22	32
Ethiopia	63,541	0.13	42.22	2.22	0.104	0.00	0.33	0.09	0.02	42.88	675	2	664	5	2	1
Fiji Islands	805	0.01	0.94	0.57	0.006	0.00	0.00	0.05	0.00	1.00	1,245	8	1,170	0	8	58
Finland	5,170	0.30	3.92	0.97	1.084	0.89	2.48	1.15	1.55	8.93	1,727	58	758	479	210	222
France	58,775	6.16	47.84	34.63	15.094	12.80	30.40	10.69	31.07	110.19	1,875	105	814	517	257	182
Gabon	1,214	0.05	1.19	0.05	0.011	0.00	0.42	0.05	0.03	1.72	1,420	43	982	347	9	38
Gambia	1,283	0.00	1.27	0.10	0.002	0.00	0.46	0.01	0.04	1.75	1,365	3	993	361	2	6
Georgia	5,271	0.72	2.66	0.52	0.534	0.19	0.22	0.03	0.05	4.17	792	137	505	42	101	6
Germany	82,169	5.45	35.64	18.84	18.771	13.15	49.59	17.50	38.48	126.95	1,545	66	434	604	228	213
Ghana	19,083	0.15	23.44	18.81	0.054	0.00	0.86	0.16	0.70	24.67	1,293	8	1,229	45	3	8
Greece	10,551	0.83	14.80	3.35	0.775	0.08	7.18	1.62	1.79	25.21	2,389	79	1,403	680	73	154
Guatemala	11,239	0.12	6.98	5.10	0.205	0.03	1.02	0.24	0.78	8.56	762	10	621	91	18	22
Guyana	759	0.02	1.46	0.99	0.011	0.00	0.07	0.04	0.05	1.60	2,113	28	1,925	93	14	54
Haiti	7,885	0.05	6.57	0.26	0.007	0.00	0.00	0.06	0.00	6.69	848	6	833	0	1	8
Honduras	6,337	0.07	4.27	2.86	0.076	0.01	0.39	0.13	0.28	4.93	778	10	673	62	12	21
Hungary	10,123	0.66	6.04	9.95	—	—	1.29	—	—	7.99	789	65	596	128	0	0
Iceland	278	0.05	0.00	0.00	0.090	0.01	0.10	0.13	0.09	0.37	1,327	183	4	348	323	470
India	1,007,369	38.62	913.70	35.29	19.065	6.04	13.75	2.24	1.24	987.38	980	38	907	14	19	2
Indonesia	204,920	5.67	236.22	22.62	0.404	0.06	26.09	1.58	2.74	269.96	1,317	28	1,153	127	2	8

(Continued)

Country	Population (1,000s)	Use of domestic water resources (10⁹ m³/yr) Domestic water withdrawal	Crop evapo-transpiration For national consumption	Crop evapo-transpiration For export	Industrial water withdrawal For national consumption	Industrial water withdrawal For export	Use of foreign water resources (10⁹ m³/yr) For national consumption Agricultural goods	For national consumption Industrial goods	For re-export of imported products	Water footprint Total (10⁹ m³/yr)	Per capita (m³/cap/yr)	Consumption of domestic water Internal water footprint	Consumption of agricultural goods Internal water footprint	External water footprint	Consumption of industrial goods Internal water footprint	External water footprint
Iran	63,202	4.68	78.58	3.18	0.984	0.60	17.90	0.51	1.03	102.65	1,624	74	1,243	283	16	8
Iraq	23,035	1.32	23.86	0.63	2.055	0.00	3.10	0.58	0.08	30.92	1,342	57	1,036	135	89	25
Israel	6,166	0.47	1.63	0.20	0.112	0.00	4.28	2.09	0.59	8.58	1,391	75	264	694	18	339
Italy	57,718	7.97	47.82	12.35	10.133	5.60	59.97	8.69	20.29	134.59	2,332	138	829	1,039	176	151
Jamaica	2,564	0.14	1.58	0.35	0.059	0.01	0.67	0.16	0.17	2.61	1,016	54	615	261	23	62
Japan	126,741	17.20	20.97	0.40	13.702	2.10	77.84	16.38	4.01	146.09	1,153	136	165	614	108	129
Jordan	4,814	0.21	1.45	0.07	0.035	0.00	4.37	0.21	0.22	6.27	1,303	44	301	908	7	43
Kazakhstan	15,192	0.59	24.87	7.92	0.035	4.58	0.29	0.06	0.33	26.96	1,774	39	1,637	19	76	4
Kenya	29,742	0.44	18.63	4.35	0.079	0.01	1.92	0.16	0.47	21.23	714	15	626	65	3	5
Korea	46,814	6.42	12.34	1.53	2.256	0.56	27.50	6.69	5.06	55.20	1,179	137	264	587	48	143
Korea, Democratic People's Republic	22,213	1.68	12.76	0.04	2.268	0.00	1.97	0.10	0.01	18.78	845	75	574	89	102	4
Kuwait	1,955	0.20	0.06	0.00	0.013	0.00	1.45	0.45	0.06	2.18	1,115	102	33	741	7	231
Kyrgyzstan	4,883	0.31	6.13	0.42	0.180	0.12	0.00	0.02	0.01	6.64	1,361	63	1,256	0	37	5
Laos	5,219	0.10	7.20	0.26	0.134	0.00	0.17	0.04	0.01	7.64	1,465	20	1,380	32	26	8
Latvia	2,383	0.16	0.70	0.32	0.073	0.02	0.55	0.14	0.30	1.63	684	67	293	232	31	61
Lebanon	4,299	0.41	1.71	0.09	0.028	0.00	3.92	0.38	0.20	6.44	1,499	95	397	913	7	88
Liberia	3,087	0.03	3.99	0.19	0.017	0.00	0.18	0.04	0.01	4.27	1,382	11	1,294	60	5	13
Libya	5,233	0.45	6.22	0.10	0.101	0.02	3.72	0.27	0.12	10.76	2,056	86	1,189	711	19	52
Lithuania	3,519	0.21	2.22	0.61	0.037	0.00	1.19	0.31	0.36	3.97	1,128	59	632	340	10	87
Madagascar	15,285	0.29	19.21	3.23	0.000	0.00	0.25	0.05	0.04	19.81	1,296	19	1,257	17	0	3
Malawi	10,205	0.12	12.71	0.77	0.037	0.00	0.13	0.03	0.01	13.03	1,277	12	1,245	13	4	3
Malaysia	22,991	1.43	36.58	18.47	0.867	0.90	12.73	2.28	8.81	53.89	2,344	62	1,591	554	38	99
Mali	10,713	0.04	21.46	3.36	0.015	0.00	0.08	0.05	0.01	21.64	2,020	3	2,003	8	1	4
Malta	390	0.04	0.05	0.00	0.000	0.00	0.47	0.17	0.02	0.75	1,916	115	141	1,212	1	448
Mauritania	2,621	0.13	2.48	0.00	0.039	0.00	0.97	0.02	0.00	3.63	1,386	48	945	369	15	10

Mauritius	1,180	0.11	0.50	0.25	0.041	0.02	0.86	0.09	0.46	1.59	1,351	91	422	729	34	75
Mexico	97,292	13.55	81.48	12.26	2.998	1.13	35.09	7.05	7.94	140.16	1,441	139	837	361	31	72
Moldova	4,284	0.24	5.15	1.65	0.765	0.86	0.13	0.03	0.07	6.31	1,474	57	1,201	31	179	6
Morocco	28,472	0.81	35.99	1.33	0.224	0.04	6.07	0.51	0.31	43.60	1,531	28	1,264	213	8	18
Mozambique	17,507	0.06	19.36	1.12	0.013	0.00	0.00	0.05	0.00	19.49	1,113	4	1,106	0	1	3
Myanmar	47,451	0.34	73.89	1.53	0.149	0.00	0.94	0.17	0.02	75.49	1,591	7	1,557	20	3	4
Namibia	1,737	0.08	0.96	0.10	0.009	0.00	0.05	0.09	0.01	1.19	683	46	555	27	5	51
Nepal	22,773	0.27	18.35	0.19	0.031	0.00	0.60	0.08	0.01	19.33	849	12	806	26	1	4
Netherlands	15,865	0.44	0.50	2.51	2.562	2.20	9.30	6.61	52.84	19.40	1,223	28	31	586	161	417
Nicaragua	5,007	0.18	3.32	1.98	0.029	0.00	0.47	0.09	0.29	4.10	819	37	663	95	6	19
Nigeria	125,375	1.41	240.38	8.54	0.383	0.30	5.59	0.31	0.43	248.07	1,979	11	1,917	45	3	2
Norway	4,474	0.45	1.09	0.17	1.032	0.43	2.42	1.57	1.02	6.56	1,467	101	244	541	231	350
Oman	2,385	0.08	0.81	0.06	0.022	0.00	2.65	0.27	0.21	3.83	1,606	32	341	1,110	9	114
Pakistan	136,476	2.88	152.75	7.57	1.706	1.28	8.55	0.33	0.67	166.22	1,218	21	1,119	63	12	2
Panama	2,832	0.50	1.57	0.67	0.034	0.00	0.47	0.20	0.20	2.77	979	178	555	165	12	69
Papua New Guinea	5,068	0.04	5.04	3.20	0.018	0.01	5.02	0.05	3.22	10.16	2,005	7	994	990	4	10
Paraguay	5,212	0.08	5.68	6.07	0.036	0.00	0.14	0.14	0.15	6.07	1,165	15	1,089	27	7	27
Peru	25,753	1.47	12.59	1.82	1.379	0.32	4.21	0.37	0.69	20.02	777	57	489	163	54	14
Philippines	75,750	4.50	99.09	7.61	0.805	1.69	11.74	0.71	2.40	116.85	1,543	59	1,308	155	11	9
Poland	38,653	1.85	21.62	2.78	6.890	4.15	10.41	1.85	2.45	42.62	1,103	48	559	269	178	48
Portugal	9,997	1.09	8.00	1.47	1.411	0.62	10.55	1.59	2.64	22.63	2,264	109	800	1,055	141	159
Qatar	573	0.07	0.11	0.01	0.008	0.00	0.24	0.19	0.02	0.62	1,087	122	196	422	14	333
Romania	22,451	2.04	29.03	2.51	3.527	4.75	3.99	0.34	0.80	38.92	1,734	91	1,293	178	157	15
Russian Federation	145,879	14.34	201.26	8.96	13.251	34.83	41.33	0.80	3.94	270.98	1,858	98	1,380	283	91	5
Rwanda	7,605	0.04	8.10	0.22	0.011	0.00	0.25	0.01	0.01	8.42	1,107	5	1,066	34	1	2
Saudi Arabia	20,504	1.61	10.42	0.42	0.181	0.01	12.11	1.58	0.62	25.90	1,263	78	508	591	9	77
Senegal	9,405	0.08	15.02	2.69	0.042	0.01	2.95	0.06	0.54	18.16	1,931	9	1,597	314	4	7
Sierra Leone	4,982	0.02	4.28	0.09	0.011	0.00	0.15	0.01	0.00	4.46	896	5	858	29	2	2
Somalia	8,627	0.02	5.05	0.15	0.001	0.00	0.71	0.01	0.02	5.79	671	2	585	82	0	1
South Africa	42,387	2.43	27.32	6.05	1.123	0.40	7.18	1.42	2.10	39.47	931	57	644	169	26	33
Spain	40,418	4.24	50.57	17.44	5.567	1.73	27.11	6.50	11.37	93.98	2,325	105	1,251	671	138	161
Sri Lanka	18,336	0.25	21.72	2.29	0.165	0.09	1.32	0.24	0.26	23.69	1,292	14	1,185	72	9	13
Sudan	30,833	0.89	66.62	7.47	0.189	0.04	0.48	0.07	0.07	68.25	2,214	29	2,161	15	6	2
Suriname	416	0.03	0.45	0.17	0.006	0.01	0.02	0.01	0.03	0.51	1,234	70	1,082	46	14	23
Swaziland	1,031	0.02	1.00	0.00	0.024	0.01	0.17	0.05	0.01	1.26	1,225	16	969	170	24	47
Sweden	8,868	1.07	4.49	1.42	1.166	0.45	4.52	3.12	2.62	14.37	1,621	121	507	509	132	352
Switzerland	7,165	0.45	0.97	0.23	1.057	0.33	5.59	3.98	2.56	12.05	1,682	63	136	780	148	555
Syria	15,994	0.59	25.40	4.08	0.246	0.08	2.82	0.16	0.50	29.22	1,827	37	1,588	176	15	10

(Continued)

Country	Population (1,000s)	Use of domestic water resources (10⁹ m³/yr)					Use of foreign water resources (10⁹ m³/yr)			Water footprint		Water footprint by consumption category (m³/cap/yr)				
		Domestic water withdrawal	Crop evapo-transpiration		Industrial water withdrawal		For national consumption		For re-export of imported products	Total (10⁹ m³/yr)	Per capita (m³/cap/yr)	Consumption of domestic water	Consumption of agricultural goods		Consumption of industrial goods	
			For national consumption	For export	For national consumption	For export	Agricultural goods	Industrial goods				Internal water footprint	Internal water footprint	External water footprint	Internal water footprint	External water footprint
Tajikistan	6,181	0.43	5.38	1.05	—	—	0.00	—	—	5.80	939	69	870	0	0	0
Tanzania	33,299	0.11	36.39	3.15	0.024	0.00	0.90	0.08	0.08	37.51	1,127	3	1,093	27	1	3
Thailand	60,488	1.83	120.17	38.49	1.239	0.55	8.73	2.49	3.90	134.46	2,223	30	1,987	144	20	41
Togo	4,457	0.07	5.28	1.82	0.011	0.00	0.30	0.03	0.11	5.69	1,277	15	1,185	68	2	7
Trinidad and Tobago	1,297	0.20	0.47	0.18	0.056	0.02	0.48	0.14	0.24	1.35	1,039	158	365	368	43	105
Tunisia	9,507	0.35	12.22	9.47	0.067	0.01	2.09	0.46	1.68	15.18	1,597	36	1,285	220	7	49
Turkey	66,850	5.38	84.05	9.81	2.731	1.07	13.68	2.11	2.43	107.95	1,615	80	1,257	205	41	32
Turkmenistan	5,184	0.38	8.39	1.07	0.118	0.05	0.18	0.07	0.05	9.14	1,764	74	1,619	36	23	13
UK	58,669	2.21	12.79	3.38	6.673	1.46	34.73	16.67	12.83	73.07	1,245	38	218	592	114	284
Ukraine	49,701	4.60	54.14	10.09	3.673	9.71	2.72	0.26	1.21	65.40	1,316	93	1,089	55	74	5
USA	280,343	60.80	334.24	138.96	170.777	44.72	74.91	55.29	45.62	696.01	2,483	217	1,192	267	609	197
Uzbekistan	24,568	2.68	18.93	6.24	1.151	0.00	1.06	0.23	0.35	24.04	979	109	771	43	47	9
Venezuela	23,938	2.80	12.42	1.28	0.360	0.13	4.86	0.70	0.76	21.14	883	117	519	203	15	29
Vietnam	78,021	3.77	85.16	11.00	11.280	0.00	2.27	0.85	0.29	103.33	1,324	48	1,091	29	145	11
Yemen	17,278	0.23	6.59	0.18	0.033	0.00	3.71	0.14	0.11	10.70	619	14	381	214	2	8
Zambia	9,980	0.28	6.94	0.51	0.058	0.07	0.22	0.03	0.05	7.52	754	28	695	22	6	3
Zimbabwe	12,497	0.21	11.48	3.35	0.085	0.02	0.00	0.12	0.03	11.90	952	17	919	0	7	10
Global total (average)	5,994,252	344	5,434	957	476	240	957	240	427	7,452	1,243	57	907	160	79	40

Appendix V

Water Footprint versus Water Scarcity, Self-Sufficiency, and Water Import Dependency per Country. Period: 1997–2001

Country	Total renewable water resources (10^9 m³/yr)	Internal water footprint (10^9 m³/yr)	External water footprint (10^9 m³/yr)	Total water footprint (10^9 m³/yr)	Water scarcity (%)	Water self-sufficiency (%)	Water import dependency (%)
Afghanistan	65.00	16.81	0.48	17.29	27	97	3
Albania	41.70	2.75	1.09	3.84	9	72	28
Algeria	14.41	24.49	12.21	36.69	255	67	33
Angola	184.00	11.49	1.52	13.00	7	88	12
Argentina	814.00	48.32	3.34	51.66	6	94	6
Armenia	10.53	2.16	0.66	2.81	27	77	23
Australia	492.00	21.77	4.80	26.56	5	82	18
Austria	77.70	4.81	8.21	13.02	17	37	63
Azerbaijan	30.28	6.51	1.32	7.83	26	83	17
Bahrain	0.12	0.16	0.61	0.77	660	20	80
Bangladesh	1,210.64	112.44	4.05	116.49	10	97	3
Barbados	0.08	0.16	0.20	0.36	445	45	55
Belarus	58.00	9.01	3.73	12.74	22	71	29
Belgium-Luxembourg	21.40	3.76	15.45	19.21	90	20	80
Belize	18.56	0.35	0.04	0.39	2	91	9
Benin	24.80	10.52	0.38	10.91	44	96	4
Bhutan	95.00	0.73	0.10	0.83	1	88	12
Bolivia	622.53	9.21	0.72	9.93	2	93	7
Botswana	14.40	0.56	0.47	1.03	7	55	45

Brazil	8,233.00	215.72	17.87	233.59	3	92	8
Bulgaria	21.30	9.91	1.42	11.33	53	87	13
Burkina Faso	12.50	16.69	0.34	17.03	136	98	2
Burundi	3.60	7.02	0.13	7.16	199	98	2
Cambodia	476.11	20.45	0.54	20.99	4	97	3
Cameroon	285.50	15.26	0.82	16.09	6	95	5
Canada	2,902.00	49.98	12.81	62.80	2	80	20
Cape Verde	0.30	0.36	0.06	0.43	142	85	15
Central African Republic	144.40	3.95	0.05	4.00	3	99	1
Chad	43.00	14.94	0.09	15.03	35	99	1
Chile	922.00	7.34	4.79	12.13	1	61	39
China	2,896.57	825.94	57.44	883.39	30	93	7
Colombia	2,132.00	28.75	5.30	34.05	2	84	16
Congo, Democratic Republic	1,283.00	36.42	0.47	36.89	3	99	1
Costa Rica	112.40	3.44	0.89	4.33	4	79	21
Côte d'Ivoire	81.00	26.98	1.09	28.06	35	96	4
Cuba	38.12	17.23	1.90	19.13	50	90	10
Cyprus	0.78	0.59	1.08	1.67	214	35	65
Czech Republic	13.15	11.44	4.70	16.15	123	71	29
Denmark	6.00	3.04	4.64	7.68	128	40	60
Dominican Republic	21.00	7.68	0.46	8.14	39	94	6

(Continued)

Country	Total renewable water resources (10^9 m³/yr)	Internal water footprint (10^9 m³/yr)	External water footprint (10^9 m³/yr)	Total water footprint (10^9 m³/yr)	Water scarcity (%)	Water self-sufficiency (%)	Water import dependency (%)
Ecuador	432.00	14.14	1.12	15.26	4	93	7
Egypt	58.30	56.37	13.13	69.50	119	81	19
El Salvador	25.25	4.10	1.31	5.41	21	76	24
Ethiopia	110.00	42.46	0.42	42.88	39	99	1
Fiji Islands	28.55	0.96	0.05	1.00	4	95	5
Finland	110.00	5.30	3.63	8.93	8	59	41
France	203.70	69.10	41.09	110.19	54	63	37
Gabon	164.00	1.26	0.47	1.72	1	73	27
Gambia	8.00	1.28	0.47	1.75	22	73	27
Georgia	63.33	3.92	0.25	4.17	7	94	6
Germany	154.00	59.86	67.09	126.95	82	47	53
Ghana	53.20	23.65	1.02	24.67	46	96	4
Greece	74.25	16.41	8.80	25.21	34	65	35
Guatemala	111.27	7.30	1.26	8.56	8	85	15
Guyana	241.00	1.49	0.11	1.60	1	93	7
Haiti	14.03	6.62	0.06	6.69	48	99	1
Honduras	95.93	4.41	0.52	4.93	5	89	11
Hungary	104.00	6.70	1.29	7.99	8	84	16
Iceland	170.00	0.14	0.23	0.37	0	38	62

India	1,896.66	971.39	15.99	987.38	52	98	2
Indonesia	2,838.00	242.30	27.66	269.96	10	90	10
Iran	137.51	84.24	18.41	102.65	75	82	18
Iraq	75.42	27.23	3.68	30.92	41	88	12
Israel	1.67	2.21	6.37	8.58	514	26	74
Italy	191.30	65.93	68.67	134.59	70	49	51
Jamaica	9.40	1.78	0.83	2.61	28	68	32
Japan	430.00	51.87	94.22	146.09	34	36	64
Jordan	0.88	1.70	4.58	6.27	713	27	73
Kazakhstan	109.61	26.60	0.35	26.96	25	99	1
Kenya	30.20	19.14	2.09	21.23	70	90	10
Korea Republic	69.70	21.02	34.18	55.20	79	38	62
Korea, Democratic People's Republic	77.14	16.70	2.07	18.78	24	89	11
Kuwait	0.02	0.28	1.90	2.18	10,895	13	87
Kyrgyzstan	20.58	6.62	0.02	6.64	32	100	0
Laos	333.55	7.44	0.21	7.64	2	97	3
Latvia	35.45	0.93	0.70	1.63	5	57	43
Lebanon	4.41	2.14	4.30	6.44	146	33	67
Liberia	232.00	4.04	0.22	4.27	2	95	5
Libya	0.60	6.77	3.99	10.76	1,793	63	37
Lithuania	24.90	2.47	1.50	3.97	16	62	38
Madagascar	337.00	19.51	0.30	19.81	6	98	2

(Continued)

Country	Total renewable water resources (10^9 m³/yr)	Internal water footprint (10^9 m³/yr)	External water footprint (10^9 m³/yr)	Total water footprint (10^9 m³/yr)	Water scarcity (%)	Water self-sufficiency (%)	Water import dependency (%)
Malawi	17.28	12.87	0.13	13.00	75	99	1
Malaysia	580.00	38.87	15.01	53.89	9	72	28
Mali	100.00	21.51	0.13	21.64	22	99	1
Malta	0.05	0.10	0.65	0.75	1,478	13	87
Mauritania	11.40	2.64	0.99	3.63	32	73	27
Mauritius	2.21	0.65	0.95	1.59	72	40	60
Mexico	457.22	98.02	42.14	140.16	31	70	30
Moldova	11.65	6.15	0.16	6.31	54	97	3
Morocco	29.00	37.02	6.58	43.60	150	85	15
Mozambique	216.11	19.43	0.05	19.49	9	100	0
Myanmar	1,045.60	74.38	1.11	75.49	7	99	1
Namibia	17.94	1.05	0.13	1.19	7	89	11
Nepal	210.20	18.66	0.68	19.33	9	96	4
Netherlands	91.00	3.50	15.91	19.40	21	18	82
Nicaragua	196.69	3.53	0.57	4.10	2	86	14
Nigeria	286.20	242.17	5.90	248.07	87	98	2
Norway	382.00	2.57	3.99	6.56	2	39	61
Oman	0.99	0.91	2.92	3.83	389	24	76
Pakistan	222.69	157.34	8.88	166.22	75	95	5

Panama	147.98	2.11	0.66	2.77	2	76	24
Papua New Guinea	801.00	5.09	5.07	10.16	1	50	50
Paraguay	336.00	5.76	0.14	5.90	2	98	2
Peru	1,913.00	15.44	4.58	20.02	1	77	23
Philippines	479.00	104.40	12.45	116.85	24	89	11
Poland	61.60	30.36	12.26	42.62	69	71	29
Portugal	68.70	10.50	12.13	22.63	33	46	54
Qatar	0.05	0.19	0.43	0.62	1,176	31	69
Romania	211.93	34.60	4.33	38.92	18	89	11
Russian Federation	4,507.25	228.85	42.13	270.98	6	84	16
Rwanda	5.20	8.15	0.27	8.42	162	97	3
Saudi Arabia	2.40	12.21	13.69	25.90	1,079	47	53
Senegal	39.40	15.14	3.02	18.16	46	83	17
Sierra Leone	160.00	4.31	0.15	4.46	3	97	3
Somalia	13.50	5.07	0.72	5.79	43	88	12
South Africa	50.00	30.87	8.60	39.47	79	78	22
Spain	111.50	60.38	33.60	93.98	84	64	36
Sri Lanka	50.00	22.13	1.56	23.69	47	93	7
Sudan	64.50	67.70	0.55	68.25	106	99	1
Suriname	122.00	0.48	0.03	0.51	0	94	6
Swaziland	4.51	1.04	0.22	1.26	28	82	18
Sweden	174.00	6.73	7.64	14.37	8	47	53
Switzerland	53.50	2.48	9.57	12.05	23	21	79

(Continued)

Country	Total renewable water resources (10^9 m³/yr)	Internal water footprint (10^9 m³/yr)	External water footprint (10^9 m³/yr)	Total water footprint (10^9 m³/yr)	Water scarcity (%)	Water self-sufficiency (%)	Water import dependency (%)
Syria	26.26	26.24	2.98	29.22	111	90	10
Tanzania	91.00	36.53	0.99	37.51	41	97	3
Thailand	409.94	123.24	11.22	134.46	33	92	8
Togo	14.70	5.36	0.33	5.69	39	94	6
Trinidad and Tobago	3.84	0.73	0.61	1.35	35	54	46
Tunisia	4.56	12.63	2.55	15.18	333	83	17
Turkey	229.30	92.16	15.79	107.95	47	85	15
Turkmenistan	24.72	8.77	0.18	8.96	36	98	2
UK	147.00	21.67	51.40	73.07	50	30	70
Ukraine	139.55	62.41	2.99	65.40	47	95	5
USA	3,069.40	565.82	130.19	696.01	23	81	19
Uzbekistan	50.41	22.76	1.29	24.04	48	95	5
Venezuela	1,233.17	15.58	5.56	21.14	2	74	26
Vietnam	891.21	100.21	3.12	103.33	12	97	3
Yemen	4.10	6.86	3.84	10.70	261	64	36
Zambia	105.20	7.27	0.25	7.52	7	97	3
Zimbabwe	20.00	11.77	0.12	11.90	59	99	1

Glossary

Virtual-water content – The virtual-water content of a product (a commodity, good, or service) is the volume of freshwater used to produce the product, measured at the place where the product was actually produced (production-site definition). It refers to the sum of the water use in the various steps of the production chain. The virtual-water content of a product can also be defined as the volume of water that would have been required to produce the product at the place where the product is consumed (consumption-site definition). If not mentioned otherwise, we use in this book the production-site definition. The adjective "virtual" refers to the fact that most of the water used to produce a product is not contained in the product. The real-water content of products is generally negligible compared with the virtual-water content.

> *The three colors of a product's virtual-water content* – The virtual-water content of a product consists of three components, namely a green, a blue, and a gray component. The "green" virtual-water content of a product is the volume of rainwater that evaporated during the production process. This is relevant mainly for agricultural products, where it refers to the total rainwater evaporation from the field during the growing period of the crop (including both transpiration by the plants and other forms of evaporation). The "blue" virtual-water content of a product is the volume of surface water or groundwater that evaporated as a result of the production of the product. In the case of crop production, the blue water content of a crop is defined as the sum of the evaporation of irrigation water from the field and the evaporation of water from irrigation canals and artificial storage reservoirs (although for practical reasons the latter component has been left out of our studies). In the cases of industrial production and domestic water supply, the blue water content of the product or service is equal to the part of the water withdrawn from ground- or surface water that evaporates and thus does not return to the system from which it came. The "gray" virtual-water content of

a product is the volume of water that becomes polluted during its production. We have quantified this by calculating the volume of water required to dilute pollutants released into the natural water system during its production process to such an extent that the quality of the ambient water remains above agreed water quality standards. The distinction between green and blue water originates from Falkenmark (2003). It is relevant to know the ratio of green to blue water use, because the impacts on the hydrological cycle are different. Both the green and blue components in the total virtual-water content of a product refer to evaporation. The gray component refers to the volume of polluted water. Evaporated water and polluted water have in common the fact that they are both "lost," i.e. in the first instance unavailable for other uses. We say "in the first instance" because evaporated water may come back as rainfall above land elsewhere and polluted water may become clean in the longer term, but these are considered here as secondary effects that will never take away the primary effects.

Virtual-water flow – The virtual-water flow between two nations or regions is the volume of virtual water that is being transferred from one place to another as a result of product trade.

Virtual-water export – The virtual-water export of a country or region is the volume of virtual water associated with the export of goods or services from the country or region. It is the total volume of water required to produce the products for export.

Virtual-water import – The virtual-water import of a country or region is the volume of virtual water associated with the import of goods or services into the country or region. It is the total volume of water used (in the export countries) to produce the products. Viewed from the perspective of the importing country, this can be seen as an additional source of water that supplements domestically available water resources.

Virtual-water balance – The virtual-water balance of a country over a certain time period is defined as the net import of virtual water over this period, which is equal to the gross import of virtual water minus the gross export. A positive virtual-water balance implies net inflow of virtual water to the country from other countries. A negative balance means net outflow of virtual water.

Water footprint – The water footprint of an individual or community is defined as the total volume of freshwater that is used to produce the goods and services consumed by the individual or community. A water footprint can be calculated for any well-defined group of consumers, including a family, business, village, city, province, state, or nation. A water footprint is generally expressed in terms of the volume of water use per year.

Water footprint of an individual – This is defined as the total amount of water used for the production of the goods and services consumed by the individual. It can be estimated by multiplying all goods and services consumed by their respective virtual-water content. A simple web-based water-footprint calculator for assessing your individual water footprint is available at www.waterfootprint.org.

Water footprint of a nation – This is defined as the total amount of water that is used to produce the goods and services consumed by the inhabitants of the nation. The national water footprint can be assessed in two ways. The bottom-up approach is to consider the sum of all goods and services consumed multiplied by their respective virtual-water content. In the top-down approach, applied in this book, the water footprint of a nation is calculated as the total use of domestic water resources plus the gross virtual-water import minus the gross virtual-water export.

Internal and external water footprint – The total water footprint of a country includes two components: the part of the footprint that falls inside the country (internal water footprint) and the part that impacts on other countries in the world (external water footprint). The distinction refers to the appropriation of domestic water resources versus the appropriation of foreign water resources.

Blue, green, and gray components of the total water footprint – The total water footprint of an individual or community breaks down into three components: the blue, green, and gray water footprints. The blue water footprint is the volume of freshwater that evaporated from the global blue water resources (surface water and groundwater) to produce the goods and services consumed by the individual or community. The green water footprint is the volume of water evaporated from the global green water resources (rainwater stored in the soil as soil moisture). The gray water footprint is the volume of polluted water associated with the production of all goods and services for the individual or community. The last has been calculated as the volume of water that is required to dilute pollutants to such an extent that the quality of the water remains above agreed water quality standards.

Water saving through trade – A nation can conserve its domestic water resources by importing a water-intensive product instead of producing it domestically. International trade can save water globally if a water-intensive commodity is traded from an area where it is produced with high water productivity (resulting in products with low virtual-water content) to an area with lower water productivity.

Water self-sufficiency versus water dependency – The "water self-sufficiency" of a nation is defined as the ratio of the internal water footprint to the total water footprint of that country. It denotes a nation's capability of supplying the water needed to meet the domestic demand for goods and services. Self-sufficiency is 100% if all the water needed is available and indeed taken from within the nation's own territory. Water self-sufficiency approaches zero if the demand for goods and services in a country is largely met with virtual-water imports. Countries with import of virtual water depend, de facto, on the water resources available in other parts of the world. The "virtual-water import dependency" of a country or region is defined as the ratio of the external water footprint of the country or region to its total water footprint.

References

Albersen, P.J., Houba, H.E.D., and Keyzer, M.A. (2003) Pricing a raindrop in a process-based model: general methodology and a case study of the Upper-Zambezi. *Physics and Chemistry of the Earth* 28: 183–192.

Allan, J.A. (1998a) Watersheds and problemsheds: Explaining the absence of armed conflict over water in the Middle East. *Middle East Review of International Affairs* 2(1): 49–51.

Allan, J.A. (1998b) Virtual water: A strategic resource, global solutions to regional deficits. *Groundwater* 36(4): 545–546.

Allan, J.A. (1999a) Water stress and global mitigation: Water, food and trade. *Arid Land Newsletter* 45.

Allan, J.A. (1999b) Productive efficiency and allocative efficiency: why better water management may not solve the problem. *Agricultural Water Management* 40: 71–75.

Allan, J.A. (2001) *The Middle East water question: Hydropolitics and the global economy.* I.B. Tauris, London.

Allan, J.A. (2003) Virtual water – the water, food, and trade nexus: Useful concept or misleading metaphor? *Water International* 28(1): 106–113.

Allen, R.G., Pereira, L.S., Raes, D., and Smith, M. (1998) *Crop evapotranspiration: Guidelines for computing crop water requirements.* FAO Irrigation and Drainage Paper 56, FAO, Rome.

Barlow, M., and Clarke, T. (2002) *Blue gold: The battle against corporate theft of the world's water.* The New Press, New York.

Berkoff, J. (2003) China: the South–North Water Transfer Project – is it justified? *Water Policy* 5: 1–28.

Bogardi, J., and Castelein, S. (eds.) (2003) Selected papers of the international conference *From Conflict to Cooperation in International Water Resources Management: Challenges and Opportunities*, 20–22 November 2002, UNESCO-IHE, Delft, The Netherlands. PCCP Publication No. 31, UNESCO, Paris.

Bressani, R. (2003) *Coffea arabica*: Coffee, coffee pulp. In: Animal Feed Resources Information System (AFRIS), Food and Agriculture Organization, Rome.

Brown, L.R. (1995) *Who will feed China? Wakeup call for a small planet.* W.W. Norton, New York.

CCI (2005) *Regions of US productions.* Cotton Council International.

Chapagain, A.K. (2006) *Globalisation of water: Opportunities and threats of virtual water trade.* PhD thesis, UNESCO-IHE, Delft.

Chapagain, A.K., and Hoekstra, A.Y. (2003) *Virtual water flows between nations in relation to trade in livestock and livestock products.* Value of Water Research Report Series No. 13, UNESCO-IHE, Delft.

Chapagain, A.K., and Hoekstra, A.Y. (2004) *Water footprints of nations.* Value of Water Research Report Series No. 16, vols. 1 and 2, UNESCO-IHE, Delft.

Chapagain, A.K., and Hoekstra, A.Y. (2007a) The global component of freshwater demand and supply: An assessment of virtual water flows between nations as a result of trade in agricultural and industrial products. *Water International,* in press.

Chapagain, A.K., and Hoekstra, A.Y. (2007b) The water footprint of coffee and tea consumption in the Netherlands. *Ecological Economics,* in press.

Chapagain, A.K., Hoekstra, A.Y., and Savenije, H.H.G. (2006a) Water saving through international trade of agricultural products. *Hydrology and Earth System Sciences* 10(3): 455–468.

Chapagain, A.K., Hoekstra, A.Y., Savenije, H.H.G., and Gautam, R. (2006b) The water footprint of cotton consumption: An assessment of the impact of worldwide consumption of cotton products on the water resources in the cotton producing countries. *Ecological Economics* 60(1): 186–203.

Cosgrove, W.J., and Rijsberman, F.R. (2000) *World water vision: Making water everybody's business.* Earthscan, London.

Cotton Australia (2005) How to grow a pair of Jeans. www.cottonaustralia. com.au.

CRC (2004) *NUTRIpak: A practical guide to cotton nutrition.* Australian Cotton Cooperative Research Centre.

Dastane, N.G. (1978) *Effective rainfall in irrigated agriculture.* FAO Irrigation and Drainage Paper No. 25, Food and Agriculture Organization, Rome.

De Fraiture, C., Cai, X., Amarasinghe, U., Rosegrant, M., and Molden, D. (2004) *Does international cereal trade save water? The impact of virtual water trade on global water use.* Comprehensive Assessment Research Report 4, IWMI, Colombo.

De Man, R. (2001) *The global cotton and textile chain: Substance flows, actors and co-operation for sustainability.* A study in the framework of WWF's Freshwater and Cotton Programme, Reinier de Man Publications, Leiden.

Donkers, H. (1994) *De witte olie: Water, vrede en duurzame ontwikkeling in het Midden-Oosten.* Uitgeverij Jan van Arkel, Utrecht.

EC Statistical Office (2006) Eurostat, Statistical Office of the European Commission, Luxembourg. http://epp.eurostat.cec.eu.int.

EPA (2005) *List of drinking water contaminants: Ground water and drinking water.* US Environmental Protection Agency.

Falkenmark, M. (2003) Freshwater as shared between society and eco-systems: from divided approaches to integrated challenges. *Philosophical Transactions of the Royal Society of London B* 358(1440): 2037–2049.

FAO (1999) *Irrigation in Asia in figures.* Water Reports 18, Food and Agriculture Organization, Rome.

FAO (2003a) *Review of world water resources by country.* Water Reports 23, Food and Agriculture Organization, Rome.

FAO (2003b) *Technical conversion factors for agricultural commodities.* Food and Agriculture Organization, Rome.

FAO (2003c) *Unlocking the water potential of agriculture.* Food and Agriculture Organization, Rome.

FAO (2006a) AQUASTAT database. Food and Agriculture Organization, Rome. www.fao.org/ag/agl/aglw/aquastat/main/index.stm.

FAO (2006b) CROPWAT decision support system. Food and Agriculture Organization, Rome. www.fao.org/ag/AGL/aglw/cropwat.stm.

FAO (2006c) CLIMWAT database. Food and Agriculture Organization, Rome. www.fao.org/ag/AGL/aglw/climwat.stm.

FAO (2006d) FAOCLIM: a CD-ROM with world-wide agroclimatic data. Food and Agriculture Organization, Rome. www.fao.org/sd/2001/EN1102_en.htm.

FAO (2006e) FAOSTAT database. Food and Agriculture Organization, Rome. http://faostat.fao.org.

Gillham, F.E.M., Bell, T.M., Arin, T., Matthews, T.A., Rumeur, C.L., and Hearn, A.B. (1995) *Cotton production prospects for the next decade.* Technical Paper No. 287, World Bank, Washington, D.C.

Glantz, M.H. (1998) Creeping environmental problems in the Aral Sea basin. In: Kobori, I., and Glantz, M.H. (eds.), *Central Eurasian water crisis: Caspian, Aral and Dead Seas.* United Nations University Press, New York.

Gleick, P.H. (ed.) (1993) *Water in crisis: A guide to the world's fresh water resources.* Oxford University Press, Oxford.

Gleick, P.H. (1998) The human right to water. *Water Policy* 1: 487–503.

Greenaway, F., Hassan, R., and Reed, G.V. (1994) An empirical-analysis of comparative advantage in Egyptian agriculture. *Applied Economics* 26(6): 649–657.

Green Cross International (2000) *National sovereignty and international watercourses.* Green Cross International, Geneva, Switzerland.

GTZ (2002a) *Post harvesting processing: Facts and figures.* PPP Project, Gesellschaft für Technische Zusammenarbeit, Germany.

GTZ (2002b) *Post harvesting processing: Coffee waste water.* PPP Project, Gesellschaft für Technische Zusammenarbeit, Germany.

GTZ (2002c) *Post harvesting processing: Coffee drying.* PPP Project, Gesellschaft für Technische Zusammenarbeit, Germany.

Haddadin, M.J. (2003) Exogenous water: A conduit to globalization of water resources. In: Hoekstra, A.Y. (ed.), *Virtual water trade: Proceedings of the International Expert Meeting on Virtual Water Trade*, pp. 159–169. Value of Water Research Report Series No. 12, UNESCO-IHE, Delft.

Haddadin, M.J. (ed.) (2006) *Water resources in Jordan: Evolving policies for development, the environment, and conflict resolution.* RFF Press.

Hall, M., Dixon, J., Gulliver, A., and Gibbon, D. (eds.) (2001) *Farming systems and poverty: Improving farmers' livelihoods in a changing world.* Food and Agriculture Organization, Rome/World Bank, Washington, D.C.

Herendeen, R.A. (2004) Energy analysis and EMERGY analysis – A comparison. *Ecological Modelling* 178: 227–237.

Hicks, P.A. (2001) Postharvest processing and quality assurance for speciality/organic coffee products. In: Chapman, K., and Subhadrabandhu, S. (eds.), *The First Asian Regional Round-table on Sustainable, Organic and Speciality Coffee Production, Processing and Marketing*, 26–28 February 2001, Chiang Mai, Thailand.

Hoekstra, A.Y. (1998) *Perspectives on water: A model-based exploration of the future.* International Books, Utrecht.

Hoekstra, A.Y. (ed.) (2003) *Virtual water trade: Proceedings of the International Expert Meeting on Virtual Water Trade*, Delft, The Netherlands, 12–13 December 2002. Value of Water Research Report Series No. 12, UNESCO-IHE, Delft.

Hoekstra, A.Y. (2006) *The global dimension of water governance: Nine reasons for global arrangements in order to cope with local water problems.* Value of Water Research Report Series No. 20, UNESCO-IHE, Delft.

Hoekstra, A.Y., and Hung, P.Q. (2002) *Virtual water trade: A quantification of virtual water flows between nations in relation to international crop trade.* Value of Water Research Report Series No. 11, UNESCO-IHE, Delft.

Hoekstra, A.Y., and Hung, P.Q. (2005) Globalisation of water resources: International virtual water flows in relation to crop trade. *Global Environmental Change* 15(1): 45–56.

Hoekstra, A.Y., and Chapagain, A.K. (2007a) Water footprints of nations: water use by people as a function of their consumption pattern. *Water Resources Management* 21(1): 35–48.

Hoekstra, A.Y., and Chapagain, A.K. (2007b) The water footprints of Morocco and the Netherlands: Global water use as a result of domestic consumption of agricultural commodities. *Ecological Economics*, in press.

Hoekstra, A.Y., Savenije, H.H.G., and Chapagain, A.K. (2001) An integrated approach towards assessing the value of water: A case study on the Zambezi basin. *Integrated Assessment* 2(4): 199–208.

Hoekstra, A.Y., Savenije, H.H.G., and Chapagain, A.K. (2003) The value of rainfall: upscaling economic benefits to the catchment scale. In: *Proceedings SIWI Seminar "Towards catchment hydrosolidarity in a world of uncertainties,"* Stockholm, August 16, 2003, pp. 63–68. Report 18, Stockholm International Water Institute, Stockholm.

ICO (2003) Website of the International Coffee Organization, London. www.ico.org.

ICWE (1992) The Dublin statement on water and sustainable development, International Conference on Water and the Environment, Dublin.

IFA, IFDC, IPI, PPI, and FAO (2002) *Fertilizer use by crop.* Food and Agriculture Organization, Rome.

ITC (2002) PC-TAS version 1995–1999, Harmonized System, CD-ROM. International Trade Centre, Geneva.

ITC (2004) PC-TAS version 1997–2001, Harmonized System, CD-ROM. International Trade Centre, Geneva.

Lechner, F.J., and Boli, J. (eds.) (2004) *The globalization reader,* 2nd edn. Blackwell Publishing, Oxford.

Loh, J., and Wackernagel, M. (2004) *Living planet report 2004.* WWF, Gland, Switzerland.

Ma, J., Hoekstra, A.Y., Wang, H., Chapagain, A.K., and Wang, D. (2006) Virtual versus real water transfers within China. *Philosophical Transactions of the Royal Society of London B* 361(1469): 835–842.

Maclean, J.L., Dawe, D.C., Hardy, B., and Hettel, G.P. (2002) *Rice almanac: Source book for the most important economic activity on earth.* International Rice Research Institute, Los Baños, Philippines.

Merrett, S. (2003) Virtual water and Occam's Razor. *Water International* 28(1): 103–105.

Mitchell, T. (2004) TYN CY 1.1. Tyndall Centre for Climate Change Research, Climatic Research Unit, University of East Anglia, UK. www.cru.uea.ac.uk/~timm/cty/obs/TYN_CY_1_1.html.

Oki, T., and Kanae, S. (2004) Virtual water trade and world water resources. *Water Science and Technology* 49(7): 203–209.

Pereira, L.S., Cordery, I., and Iacovides, I. (2002) *Coping with water scarcity.* International Hydrological Programme, UNESCO, Paris.

Postel, S. (1992) *Last oasis: Facing water scarcity.* W.W. Norton, New York.

Postel, S.L., Daily, G.C., and Ehrlich, P.R. (1996) Human appropriation of renewable fresh water. *Science* 271: 785–788.

Proto, M., Supino, S., and Malandrino, O. (2000) Cotton: a flow cycle to exploit. *Industrial Crops and Products* 11(2–3): 173–178.

Qian, Z.Y., Lin, B.N., Zhang, W.Z., and Sun, X.T. (2002) *Comprehensive report of strategy on water resources for China's sustainable development*, 1st edn, pp. 38–41. China Water and Hydropower Press, Beijing.

Ramirez-Vallejo, J., and Rogers, P. (2004) Virtual water flows and trade liberalization. *Water Science and Technology* 49(7): 25–32.

Rees, W.E. (1992) Ecological footprints and appropriated carrying capacity: what urban economics leaves out. *Environment and Urbanization* 4(2): 121–130.

Ren, X. (2000) Development of environmental performance indicators for textile process and product. *Journal of Cleaner Production* 8(6): 473–481.

Renault, D. (2003) Value of virtual water in food: Principles and virtues. In: Hoekstra, A.Y. (ed.), *Virtual water trade: Proceedings of the International Expert Meeting on Virtual Water Trade*. Value of Water Research Report Series No. 12, UNESCO-IHE, Delft.

Rosenblatt, L., Meyer, J., and Beckmann, E. (2003) *Koffie: Geschiedenis, teelt, veredeling, met 60 heerlijke koffierecepten*. Fontaine Uitgevers, Abcoude, The Netherlands.

Roast and Post (2003) *From tree to cup, processing*. The Roast and Post Coffee Company, UK.

Twinings (2003) *Black tea manufacture*. R. Twining and Company Ltd., London.

Salman, S.M.A., and McInerney-Lankford, S. (2004) *The human right to water: Legal and policy dimensions*. World Bank, Washington, D.C.

Serageldin, I. (1995) Water resources management: a new policy for a sustainable future. *Water International* 20(1): 15–21.

Seuring, S. (2004) Integrated chain management and supply chain management: Comparative analysis and illustrative cases. *Journal of Cleaner Production* 12: 1059–1071.

Shiklomanov, I.A. (2000) Appraisal and assessment of world water resources. *Water International* 25(1): 11–32.

Silvertooth, J.C., Navarro, J.C., Norton, E.R., and Galadima, A. (2001) *Soil and plant recovery of labeled fertilizer nitrogen in irrigated cotton*. Arizona Cotton Report, University of Arizona.

Smakhtin, V., Revenga, C., and Döll, P. (2004) *Taking into account environmental water requirements in global-scale water resources assessments*. Comprehensive Assessment Research Report 2, IWMI, Colombo.

Soth, J., Grasser, C., and Salerno, R. (1999) *The impact of cotton on fresh water resources and ecosystems: A preliminary analysis*. WWF, Gland, Switzerland.

UNCTAD (2005a) Planting and harvesting times for cotton, by producing country. http://r0.unctad.org/infocomm/anglais/cotton/crop.htm.

UNCTAD (2005b) Cotton uses. http://r0.unctad.org/infocomm/anglais/cotton/uses.htm.

UNEP (2002) *Global environment outlook 3: Past, present and future perspectives.* Earthscan, London.

UNEP IE (1996) *Cleaner production in textile wet processing: a workbook for trainers.* United Nations Environment Programme: Industry and Environment, Paris.

UNESCO (2003) *Water for people, water for life: The United Nations world water development report.* UNESCO Publishing, Paris / Berghahn Books, Oxford.

UNESCO (2006) *Water, a shared responsibility: The United Nations world water development report 2.* UNESCO Publishing, Paris / Berghahn Books, Oxford.

USDA (2004) Cotton: World markets and trade. www.fas.usda.gov/cotton/circular/2004/07/CottonWMT.pdf.

USDA/NOAA (2005a) Major world crop areas and climatic profiles. USDA/NOAA Joint Agricultural Weather Facility. www.usda.gov/agency/oce/waob/mississippi/MajorWorldCropAreas.pdf.

USDA/NOAA (2005b) Cotton – World supply and demand summary. USDA/NOAA Joint Agricultural Weather Facility. www.tradefutures.cc/education/cotton/worldsd.htm.

USEPA (1996) Best management practices for pollution prevention in the textile industry. www.e-textile.org.

Van den Bergh, J.C.J.M., and Verbruggen, H. (1999) Spatial sustainability, trade and indicators: An evaluation of the 'ecological footprint'. *Ecological Economics* 29: 61–72.

Van Kooten, G.C., and Bulte, E.H. (2000) The ecological footprint: useful science or politics. *Ecological Economics* 32: 385–389.

Visvanathan, C., Kumar, S., and Han, S. (2000) Cleaner production in textile sector: Asian scenario. Paper presented at the *National Workshop on Sustainable Industrial Development through Cleaner Production*, 12–13 November, Colombo.

Vörösmarty, C.J., and Sahagian, D. (2000) Anthropogenic disturbance of the terrestrial water cycle. *BioScience* 50(9): 753–765.

Vörösmarty, C.J., Green, P., Salisbury, J., and Lammers, R.B. (2000) Global water resources: Vulnerability from climate change and population growth. *Science* 289: 284–288.

Wackernagel, M., and Rees, W. (1996) *Our ecological footprint: Reducing human impact on the earth.* New Society Publishers, Gabriola Island, B.C., Canada.

Wackernagel, M., Onisto, L., Linares, A.C., Falfan, I.S.L., Garcia, J.M., Guerrero, I.S., and Guerrero, M.G.S. (1997) *Ecological footprints of nations: How much nature do they use? – How much nature do they have?* Centre for Sustainability Studies, Universidad Anahuac de Xalapa, Mexico.

Wackernagel, M., Onisto, L., Bello, P., Linares, A.C., Falfan, I.S.L., Garcia, J.M., Guerrero, A.I.S., and Guerrero, M.G.S. (1999) National natural capital accounting with the ecological footprint concept. *Ecological Economics* 29: 375–390.

WHO (2003) *The right to water.* World Health Organization, Geneva.

Wichelns, D. (2001) The role of 'virtual water' in efforts to achieve food security and other national goals, with an example from Egypt. *Agricultural Water Management* 49(2): 131–151.

Wichelns, D. (2004) The policy relevance of virtual water can be enhanced by considering comparative advantages. *Agricultural Water Management* 66(1): 49–63.

World Bank (1999) *Pollution prevention and abatement handbook 1998: Toward cleaner production.* World Bank, Washington, D.C.

World Bank (2004) *World development indicator data-query.* World Bank, Washington, D.C.

WTO (2004) Statistical database. World Trade Organization, Geneva.

WWC (2004) *E-Conference Synthesis: Virtual Water Trade – Conscious Choices.* WWC Publication No. 2, World Water Council, Marseille, France.

WWF (2003) *Thirsty crops: Our food and clothes: Eating up nature and wearing out the environment? Living waters: conserving the source of life.* WWF, Zeist, The Netherlands.

WWI (2006) *State of the world 2006: Special focus: China and India.* Worldwatch Institute / W.W. Norton, New York.

Yang, H., and Zehnder, A.J.B. (2002) Water scarcity and food import: A case study for Southern Mediterranean countries. *World Development* 30(8): 1413–1430.

Yang, H., Reichert, P., Abbaspour, K.C., and Zehnder, A.J.B. (2003) A water resources threshold and its implications for food security. In: Hoekstra, A.Y. (ed.), *Virtual water trade: Proceedings of the International Expert Meeting on Virtual Water Trade*, pp. 111–117. Value of Water Research Report Series No. 12, UNESCO-IHE, Delft.

Yang, H., Wang, L., Abbaspour, K.C., and Zehnder, A.J.B. (2006) Virtual water trade: an assessment of water use efficiency in the international food trade. *Hydrology and Earth System Sciences* 10: 443–454.

Zimmer, D., and Renault, D. (2003) Virtual water in food production and global trade: Review of methodological issues and preliminary results. In: Hoekstra, A.Y. (ed.), *Virtual water trade: Proceedings of the International Expert Meeting on Virtual Water Trade*, pp. 93–109. Value of Water Research Report Series No. 12, UNESCO-IHE, Delft.

Index